Critical Semiotics

BLOOMSBURY ADVANCES IN SEMIOTICS

Semiotics has complemented linguistics by expanding its scope beyond the phoneme and the sentence to include texts and discourse, and their rhetorical, performative, and ideological functions. It has brought into focus the multimodality of human communication. *Advances in Semiotics* publishes original works in the field demonstrating robust scholarship, intellectual creativity, and clarity of exposition. These works apply semiotic approaches to linguistics and non-verbal productions, social institutions and discourses, embodied cognition and communication, and the new virtual realities that have been ushered in by the Internet. It also is inclusive of publications in relevant domains such as socio-semiotics, evolutionary semiotics, game theory, cultural and literary studies, human-computer interactions, and the challenging new dimensions of human networking afforded by social websites.

Series Editor: Paul Bouissac is Professor Emeritus at the University of Toronto (Victoria College), Canada. He is a world-renowned figure in semiotics and a pioneer of circus studies. He runs the SemiotiX Bulletin [www.semioticon.com/semiotix], which has a global readership.

TITLES IN THE SERIES:

A Buddhist Theory of Semiotics, Fabio Rambelli
Computable Bodies, Josh Berson
Introduction to Peircean Visual Semiotics, Tony Jappy
Semiotics and Pragmatics of Stage Improvisation, Domenico Pietropaolo
Semiotics of Drink and Drinking, Paul Manning
Semiotics of Happiness, Ashley Frawley
Semiotics of Religion, Robert Yelle
The Language of War Monuments, David Machin and Gill Abousnnouga
The Semiotics of Clowns and Clowning, Paul Bouissac
The Semiotics of Che Guevara, Maria-Carolina Cambre
The Visual Language of Comics, Neil Cohn

Critical Semiotics

Theory, from Information to Affect

GARY GENOSKO

Bloomsbury Academic
An imprint of Bloomsbury Publishing Plc

B L O O M S B U R Y
LONDON • OXFORD • NEW YORK • NEW DELHI • SYDNEY

Bloomsbury Academic
An imprint of Bloomsbury Publishing Plc

50 Bedford Square
London
WC1B 3DP
UK

1385 Broadway
New York
NY 10018
USA

www.bloomsbury.com

BLOOMSBURY and the Diana logo are trademarks of Bloomsbury Publishing Plc

First published 2016

British Library Cataloguing-in-Publication Data
A catalogue record for this book is available from the British Library.

ISBN: HB: 978-1-4725-9637-6
PB: 978-1-4725-9636-9
ePDF: 978-1-4725-9639-0
ePub: 978-1-4725-9638-3

Library of Congress Cataloging-in-Publication Data
Names: Genosko, Gary, author.
Title: Critical semiotics : theory, from information to affect / Gary Genosko.
Description: London ; New York : Bloomsbury Academic,
An imprint of Bloomsbury Publishing Plc, [2016] | Series: Bloomsbury Advances in Semiotics |
Includes bibliographical references.
Identifiers: LCCN 2016012287 (print) | LCCN 2016024987 (ebook) |
ISBN 9781472596376 (hardback) | ISBN 9781472596369 (pbk.) | ISBN 9781472596390 (ePDF) |
ISBN 9781472596383 (ePub) | ISBN 9781472596390 (epdf) | ISBN 9781472596383 (epub)
Subjects: LCSH: Semiotics–Social aspects. | Semantics (Philosophy) |
Communication models. | Culture–Semiotic models. | System theory. |
Iconicity (Linguistics) | Sociolinguistics.
Classification: LCC P99.4.S62 G36 2016 (print) | LCC P99.4.S62 (ebook) |
DDC 401/.4–dc23
LC record available at https://lccn.loc.gov/2016012287

Cover images: fluidworkshop / Shutterstock

Series: Bloomsbury Advances in Semiotics

Typeset by Integra Software Services Pvt. Ltd.

Contents

Introduction

Semiotics investigates meaning-making. Meaning is found in a variety of both forms and substances, which manifest it either potentially or actually, in extractable qualities and quantities. Variations of meaning considerably complicate the quests undertaken by semioticians and direct them into different worlds, human and nonhuman, where they hope to recover the codes in which meaning is nestled and that make communication possible. Semioticians hope to lay bare the elementary and more advanced structures that inform signification; they want to erect the conceptual scaffolding that enables them to analyze systems of signs and semiotic processuality. If semiotics has a "Holy Grail," it is surely meaning, and how it is generated, articulated, and later manipulated; how this can be explained and reflected upon, that is, theorized, and then modeled, is perhaps the holiest vow that can be taken in the institution of sign study. A properly semiotic investigation into meaning ends up on the side of how it is produced, and what sort of systems are involved; what kind of mediations need to be taken into account, and how to account for the relationalities involved, not to mention the veils that have to be lifted in the process. Critical semiotics begins at the moment when the *how* question is displaced by the *why* question, following the suggestion of Tony Schirato (1998: 397) in his encyclopedia entry on "meaning" in the *Encyclopedia of Semiotics*: meaning then becomes ideological. Further displacements follow: meaning differs from itself; the production of meaning fails; meaning is evacuated. As the late semiotician Scott Simpkins (2001: 3) once put it, critical semiotic thought requires an "asymmetrical approach" to the analysis of the production of meaning in semiotic theory that may, in fact, "go nowhere," but that is the point of its disruption of the status quo. It is not all about disruption. Simpkins's version of critical semiotics investigated the implications of a semiosis focused on intensity, radical openness, play, surprise, and suddenness: an unbound semiosis. He developed a critical semiotic method of inquiry and his resistance to and suspicion about pronouncements made in the name of post-semiotics was a model of scholarly probity to the extent that it reoriented skeptics from using the post- as an alibi for forgetfulness, shifting instead toward a search for perturbations in semiotic programming. This book begins with a first chapter on a-signifying semiotics and the prospects of building a hybrid semiotics that is not centered by meaning; not centered, in short, by the linguistic signifier to which meaning is in a relation of subjection. By the final chapter of this book, the idea that some vehicle conveys meanings or messages will not be denied but, instead, integrated into another approach based on the production and circulation of affects. Affect's alleged displacement of meaning carries with it a shift from isolatable messages to transformative powers, from speakers to a-subjective enunciations, and treats signs in relation to intensities that have the capacity to disrupt and open experience. These wide-ranging prospects are positive opportunities for theoretical innovation within critical semiotic thought. "Language is filled with affects," Alan Bourassa (2002: 65) once remarked in the spirit of Gilles Deleuze's philosophy, and the implication is that signs become something more than themselves when they "take up" affective intensities. The struggle here is between the implications of the extent to which affect is immanent to language and how much more affect can be taken up, reread or regained by language without capturing and arresting the flows in the forms and substances of signs and systems.

Beginning with a-signification, Félix Guattari's hybrid semiotics in Chapter 1 will orient us toward modern and high-tech examples of how money is accessed, the computational codes and other systems that make banking "convenient." However, Guattari approaches semiotics under a broad Foucauldian rubric of microphysics of force relations—not, then, of power, but of a micropolitics of change at all scales of life. He is undoubtedly influenced by the pioneering psychedelic philosopher and psychologist Timothy Leary (1968) in whose book from the 1960s, *The Politics of Ecstasy*, the term "molecular revolution" is coined with regard to fundamental change through chemical modifications of consciousness through the ingestion of LSD. Many of Guattari's readers take note of his mixing of molecules, both literal and figurative. A revolution is molecular when it proceeds by decentered, micro-scale transformational changes, while remaining communicable at global scales through machinic processes like the Internet. Without the overt drug and big "R" revolutionary debts to both Leary and Marxist-Leninism, Guattari's approach is anticapitalist but with a trenchant semiotic focus on capitalism's mutations. Chapter 1 exposes in detail the importance of molecularization as a way to break down signs into hypothetical objects known as particle-signs. One of the roles of a-signifying semiotics is to erode and decay dominant semiological conceptions of signification and release what might be called *sign dust* as the edifice frays and perhaps tears apart. This strategy gets us thinking about what Eugene Thacker (2011) calls a "world without us," a world beyond human subjecthood and meaning-making, a world of proto-enunciation, animistic powers, and even magic. For Guattari, this is a world defined by complex interactions on a vast machinic phylum that has spread over the entire planet, as well as by analogy, processes of decomposition and regeneration of plant matter by microorganisms. What is prescient is Guattari's positive post-humanist vision of human–animal–machine assemblages and the suggestion that sign dust, long before dust and particulate became important points of reference in contemporary theory-building, warrant critical attention.

It is by linking Guattari's thought with the tradition of communication modeling initiated in the postwar years by Claude Shannon and Warren Weaver that the extent to which critical semiotics rests on the displacement of meaning may be advanced. Meaning is not irrelevant, and its periodic refluxes, decenterings, and speculative theories about the vacuoles of nonmeaning that inhabit it bring into focus the contributions of Roland Barthes who, while representing a tradition grounded in a throughgoing linguistico-semiology, also attempted to theorize signifiers without signifieds. His obtuse sign hides its meaning and eludes fullness; it out-maneuvers meaning to achieve inarticulateness, which may still be surprisingly effectual.

Chapter 1 also includes a discussion of Guattari's debts to Lewis Mumford's concept of the megamachine, but recontextualized from its imperial origins and reconceptualized through an important distinction between social subjection and machinic enslavement. This allows Guattari to surmount Mumford's humanism and anti-machine rhetoric, and to fully eclipse any lingering antiurban pastoralism of the psychedelic heritage of molecular politics. It also serves as an example of how Guattari conceptualizes machinic connectivity, and sits in contrast with the smaller scale example of Guattarian machinism that I deploy of a bank card with a magnetic stripe to examine what roles digital media may play and the consequences of this kind of engagement for processes of subjectivation. Recourse to digital examples and technomateriality helps to dispel any suggestion that a-signifying semiotics is a code word for either "instantaneist "or "spontaneist" modes of communication (Guattari 2011b).

From an antilinguistics and general critique of structuralism that drives Chapter 1, Chapter 2 turns to another kind of anti-semiology in the hands of Jean Baudrillard's work from the early to mid-1970s. Baudrillard lays bare, in an uncompromising matrix of relations and transformations, the four logics of signification and the passages between their theories of value and the objects expressing them. Baudrillard's general conversion table is then reviewed, and the ways it maps, matches, and excludes certain combinatorial possibilities between use value, economic exchange value, sign exchange value, and symbolic exchange are discussed in depth. Baudrillard exposes the power imbalances between signifier and signified and denotation and connotation, while working through a number of homologies between signs and commodities. Ultimately, he wants to distinguish between all theories of value and symbolic exchange, his anthropologically derived alternative to simulation. I present a detailed example of how simulation functions (butter vs. margarine), as well as probe the anthropological foundations of Baudrillard's symbolic through his interpretation of Marcel Mauss's theory of the gift, which is filtered by the theory of general economy advanced by Georges Bataille. The concluding discussion in this chapter addresses the radicality of Baudrillard's version of Mauss in terms of the telemorphosis of reality by "reality" television.

The first two chapters embed the mutual influences between Baudrillard, Barthes, and Guattari on a number of conceptual crossovers—obtuse signs, mythologies and rhetorics of products, staggered signifying systems, denotation and connotation, sign-referent relations, and blended sign theories. These crossovers take place on hostile ground as Barthes remained wedded to a translinguistic semiology that is at odds with both Baudrillard's and Guattari's positions; if one wants to seek peace it would require turning to the shared interest in photography between Baudrillard and Guattari and how Barthes's theorization of the punctum works across these three different scenes of writing, a matter to which I will turn briefly in this book's conclusion. Barthes serves, in the end, whether as resource or foil, as a valuable point of contact for critical semiotic thought.

Critical semiotic theory orbits around Ferdinand de Saussure's *Course in General Linguistics*. Saussure's postulate that the study of language as a social institution would give rise to "*a science that studies the life of signs in society... semiology*" (2011: 16) situates linguistics within semiology and semiology remains open to classification. Thus, Saussure did not claim that the semiological problem was linguistic, but that the language problem was semiological and therefore language may be displaced from primary consideration in relation to other signifying systems. Saussure's students did not agree on his precise formulation of the language–semiology relation in compiling the notes that make up the *Course*, but even if language is the "master pattern" for semiology, this does not require all semiology to be linguistical, as linguistics is a division of semiology, and the entire project of sign theory from the ancients to the moderns is not adequately addressed by Saussure. Guattari's rejection of linguistic imperialism is a salvo directed at all of structuralism operating in the shadow of linguistics and subject to the overcoding despotism of the linguistic signifier. The philosophical countertask becomes how to think about difference in positive terms and in a way that does not in advance determine outcomes through the combinatorial possibilities given by the system. While Guattari was intent upon releasing signs from the hold of a "trendy" form of analysis with a weak purchase on scientific objectivity, because semiology avoided social and political problems, he advanced into pragmatics and semiotics since they displayed the resilience to survive a number of takeover attempts by linguistics. Guattari (2011a: 43) sought to release semiosis from the influence of dominant linguistic significations that inhibit creativity, standardize, and neutralize. All of the deviations from Saussure that will be addressed in this book taken together simply underline the role of the *Course* as a kind of irreplaceable index for critical semiotic innovation.

In Chapter 3, I develop an analysis of semiocapitalism from two perspectives. The first concerns its qualities such as immateriality, coagulation of capital-fluxes, and the coalescence of resistance to it, and, second, the semiotic problem of substance. An info-commodity is produced between immaterial and material forms of labor, often in a networked, collaborative fashion, with multiple, often complex technical knowledges widely diffused throughout an educated population (heavily focused on IT), either formally or informally, but based upon personal attributes of a worker whose work and free time tend to blend together. The mutual imbrication of immateriality and materiality of the post-Fordist division of labor is recapitulated in the products of such labor, suspending the semiocommodity in-between the producers and consumers who together make it, producing affectively rich relationships. This places cooperation inside of immaterial labor itself and potentially provides for coalescences of resistance to semiocapitalist exploitation.

The second perspective involves the ancient metaphysical problem of the formation of matter that Aristotle reflected upon in his *Metaphysics* (McKeon edition 1941) and is known as hylomorphism: the imposition of form onto a passive matter that creates a substance. In *A Thousand Plateaus*, Guattari and Deleuze (1987: 408–10) insert part-forms and a lively "energetic materiality" into matter in order to give to it an active dimension and to insist not on the imposition of forms but on following and "surrendering to" matter's fluxes. They disrupt the two thresholds of form and matter by deformation and spillage, returning to the very example Aristotle used, metal (brazen sphere), to illustrate against the grain that the continuous variability and vitality of metal belies the fixity typical of such a form and matter.

The philosophical and semiotic perspectives on hylomorphism blend together in the figure of Danish linguist Louis Hjelmslev. Semiotically formed substance is the result of the projection of form onto matter; in Hjelmslev's famous example in his *Prolegomena to a Theory of Language*, the form of a net is projected onto an undivided surface (unformed matter like clouds, sand, and thought) and the shadow is the substance, which is not as robust as the shaping of matter by form in the philosophical tradition noted earlier, but illustrates within the glossematic system that form is a constant and substance a variable (it can be formed graphically or phonologically): "substance is thus not a necessary presupposition for linguistic form, but linguistic form is a necessary presupposition for substance" (1969: 106). Hjelmslev rehearses Saussure's (2011: 113) focus on form over substance, and "differences in formation" among languages of the amorphous "non-linguistic stuff" (1969: 77) are analyzable but the "stuff" is not, and it falls to other sciences to undertake such analysis, with the proviso that they will find their own forms (nonlinguistic) in so doing. The "metaphysical" concept of substance, as Hjelmslev refers to it (1969: 81), is not of great interest to linguistic forms conceived of in "algebraic" terms as the "general calculus" of virtualities without any current actualizations. Hjelmslev's approach is more radical than Saussure's to the extent that the latter negatively excluded the description of sounds as "substance" and the former speculated that "substance" may not be necessary to the calculus he sought to establish. The search for pure differences among forms is rejected outright by Deleuze and Guattari, while reserving praise for the byzantine complexity and abstraction of Hjelmslev's ideas of potential manifestations of forms. Thus, in Chapter 3, swerving around substance and avoiding hylomorphism with Hjelmselv's assistance, that is, the designation and description of assemblages of semiotic and capital fluxes, are required in order to understand the info-commodity as an immaterial product. Guattari defined a-signifying semiotics by skirting around semiotically formed substance so that none of the intensive multiplicities were lost in being controlled by the formalized pairs of signifier–signified or expression and content; further, form, too, is deformed, rendered vague and unfixed, and as connectivity without constraints. In other words, info-commodities can be recognized without manifesting in substances, hence the recourse to soft actualizations using the suggestive term *coagulation*, purer fluxes independent of substances and forms and not frozen by them. In this way, the study of semiocapitalism enriches the semiotic lexicon, and the agile critical operations of swerving and thwarting, softening and positive partial formation.

In Chapter 4, I turn to Michel Foucault's influential book *Discipline and Punish* (1977) in order to extract from it his explication of *obstacle-signs*. Reflecting on late eighteenth-century efforts to reform the criminal justice system in France, and the double transition in crimes from violent encounters to property- and fraud-related offenses, as well as the loss of intensity as punishments became more refined, Foucault examines the reconfiguration of punishment as a future-leaning art of effects. These effects were obstacle-signs that would not only totally codify the relationship between crimes and their punishments, but widely communicate these configurations of the disadvantages of criminal behavior by means of a wide variety of public representations. The nonarbitrary, in fact, the as-far-from-arbitrary-as-possible relations, between crimes and punishments provides a rich example of a historical effort, with contemporary resonances in the messages that circulate about the consequences of drinking and driving, of how to creatively surmount the principle of the arbitrariness of the sign. The consequences of this maneuver are, however, tied to a conservative and highly reactionary politics aimed at the figure of the "drunk driver" that gained traction in the 1980s in North America and remains with us today, even though it failed historically in the rise of a single signified punishment (imprisonment) that limited its analogical diversity. The analogical or motivated relation between signifier and signified excavated by Foucault provides less an element of unpredictability than a nonabsolute language of contrast between arbitrariness and nonarbitrariness, a way of entering into analogical thought and the service that obstacle-signs provide in the arresting of social desire and the polymorphous pleasures of illegalities. Foucault stages, then, an encounter between theaters of representation (obstacle-signs) and factories of desire (ennobled crimes and common illegal pleasures like "having one [final alcoholic drink] for the road [drive home]"). For critical semiotic theory, the primary characteristic of obstacle-signs is that they are motivated, and their secondary characteristic is that they are forces of antidesire—they block pleasure in their didactic insistence.

Chapter 5 poses a vexing problem: can signs conduct intensities? If so, how? If not, why not? Between Jacques Derrida and Jean-François Lyotard, there is a shared problem of how to regain the other of signification—force, intensity, desire. One of the effects of structure is to arrest force, to pin it onto intention, to stratify and dematerialize, to elevate the signifier and subordinate the signified (hierarchize), to generate binaries and double articulations. From the core philosophies of post-structuralism to the emerging philosophy of affect, the problem of how to get change and movement back into signification and structure tends to result in aporias. Post-structuralist thought bequeaths a number of paradoxical solutions to this problem, striking in their rhetorical trappings, that show it would be incorrect to assume that some flux did not already exist in the semiological field of immanence in the first place. Lyotard's analyses of Saussure's linguistics reveal a number of options in this regard, which are linked to "the figural" as disruptive of the flat space of the system of language and the rules that govern it. I will discuss both his concept of thickening and his example of designation. Further, Lyotard's use of proper names will illustrate how he wants to bring intensity to the sign itself, in the same place that typically desire is thought not to flow, neither by reinventing the sign, nor by finding a new site of semiosis, but by having it both ways: a tensor sign that is productive of meaning *and* intensity at the same time and place. In *Anti-Oedipus*, Deleuze and Guattari (1977: 243) praise Lyotard for mounting "the first generalized critique of the signifier," though its effects were not long-lasting. Lyotard seems to have expressed some doubt about the efficacity of this position, yet he readily fell into line with the philosophy of desire that valorized the connective tissues of drives and investments ("and ... and ... and") in the spirit of Deleuze and Guattari's connective synthesis and its productive couplings, slicings, and grafts of desire. An equally valid and productive reference point would be Lyotard's interest in Freud's account in *The Interpretation of Dreams* (2001) of the operations of the dream-work, namely, displacement as the means of distortion in dreams in the change of emphasis (i.e., strength) of intensities attached to ideas, if we are to understand how the thinking of intensities gets underway in the jostling of force and meaning in the field of psychoanalysis. Chapter 5 moves toward the emergence of affect theory, which levels the charge that structure equals a kind of stasis that cannot be readily reanimated, by reinserting otherwise marginalized post-structuralist reflections on tensors into the discussion of signifying intensities. The question that this chapter ultimately grapples with is whether or not the prospect of reanimation through the tensorial can generate anything more than a *zombie sign* in a state of continual agitation at the expense of meaningfulness. Or, rather, the question is poorly posed as the sign's conduction of its "other" does not come from an outside but is, instead, already there. Can sign theory build a set of concepts and relations that would express affect's immanence without inflicting damage on it by fixing (i.e., in negative interdependent differences among linguistic signs that stabilize the powers to affect and to be affected between significant units) and granting to it a sterile virtuality (i.e., in the relative indefiniteness of associative constellations)?

In Chapter 6, the question of the conduction of intensity by signifiers glowing with a peculiar vividness is embedded in psychosis, and I consider Fredric Jameson's pathologization of the eclipse of the centered ego of the modernist subject in the scattered and therefore impersonal intensities of postmodernity. Jameson uses the linguistic signifier, albeit one broken from its chain because it has lost its interdependencies with other signifiers, to make his point about the mysteriousness of the kind of affect he wants to evoke. He presents, albeit in an indirect way that relies on Jacques Lacan's structural psychoanalysis, the torn signifier as an affect accumulator vivid in its materiality. The stranded signifier is a tool that critical semiotic theory can use to advance its inquiry into the conditions of signification by regaining materiality and intensity in a situation in which incompleteness and disability are not automatic deficits. Building up a toolbox ready for the construction of a semiotic whose signs conduct and display forces of their own, I then turn to the floating signifier and signified as they pass from the structuralism of Claude Lévi-Strauss through the semiolinguistics of Barthes into the speculative libidinal economics of Lyotard. The imbalance between a signifier surfeit and deficit of signifieds, that Lévi-Strauss theorizes through the role of *mana* in Marcel Mauss's theory of the gift, is managed by symbolic thought in the name of *flotation*: a symbol in a pure state of emptiness ("zero symbolic value") with the capacity to absorb any meaning that would bring balance (cosmological) to the signifier–signified relationship. Although flotation is used in postmodern theory as an index of anchorlessness and indeterminate drifting, Lévi-Strauss shows how useful such looseness can be for magical thinking that seeks to provide a description of the universe in terms of an abundant resource that needs to be managed. Such magical thinking, reserved by ethnographers for the epistemologies of indigenous peoples, is not unknown among sign theorists as Barthes (1982) makes evident in his management of signifying surpluses in his study, *Empire of Signs*, of a "fictive" nation, Japan. Diverging from the approach initiated by

Brian Rotman (1987) in his analysis of simulated money (xenomoney) as nonreferential and self-reflexive in *Signifying Nothing*, and his partial maintenance of the mathematical origin story of zero against deconstruction, the question of the efficacity of zero as a semiotic construct is posed in terms of how a fillable emptiness provides a robust site for problem-solving. Indeed, the application of such a zero exceeds in significance the outcome of multiplying any natural number by zero.

1

From Information Theory to Félix Guattari's A-signifying Semiotics

The fundamental problem of communication is that of reproducing at one point either exactly or approximately a message selected at another point. Frequently the messages have meaning*; that is they refer to or are correlated according to some system with certain physical or conceptual entities. These semantic aspects of communication are irrelevant to the engineering problem. The significant aspect is that the actual message is one selected from a set of possible messages.*

(SHANNON 1948: 379)

A-signification, understood in its most general sense as any semiotic that dissociates itself in some manner from a meaning component, or considers meaning to be an irritant, has an approximate birthdate in the postwar years of the late 1940s. The moment when information theorist Claude Shannon contrasted an everyday definition of information based on semantic content with a technical one based on uncertainty, the so-called irrelevance of meaning for communication understood as an engineering problem was born. This bold move away from content did not hold for very long. Shannon's colleague, Cold War bureaucrat of big science Warren Weaver, worked the "semantic problem" back into his popular explanation of Shannon's communication model shortly thereafter. Once out of the bottle, however, the genie of meaning has had to run an obstacle course against the forces and factors displacing it; that is to say, recourse to what could be communicated, defined logarithmically in bits, the probabilistics of choice, and the redundancies that shape it, all of which determine the relative entropy of theoretical information systems. Shannon's interests in both abstract and concrete mathematical machines, especially relay circuitry and secrecy systems, but also chess-playing computers and electromechanical maze-solving mice, appear to open a proto-machinic perspective in the spirit of Guattarian thought.

But not so fast. While it seems obvious to index an "origin tale" for a-signifying semiotics on postwar information theory since it provides an influential example of expunging meaning from the foundational model of communication—or what I call the model "to-," de rigueur index of all such modeling (Genosko 2012: 115)—Guattari would have seen this as merely an episodic "skirmish" with meaning. After all, Weaver (1949: 96) justified consideration of the semantic problem of communication only to the extent that it is overlapped by the theory of the technical problem, namely accuracy. Weaver was focused on understanding the receiver of messages and his or her behaviors. The real issue remained, quite clearly, the extent to which primary, technical, a-signifying messaging overlapped and subsumed analytic, secondary, and tertiary levels of meaning and effectiveness (those affecting conduct).

Guattari (1977: 335) regarded information theory's "skirmish" with meaning as a "rearguard semiological conflict"—without mentioning Weaver specifically. What Weaver does is add new stations to the communication model of Shannon, even if, at the same time, these stations capture and arrest destratifying tendencies from Shannon's initial eschewing of meaning. Weaver increases the number of boxes/stations within the model of communication by depositing a semantic receiver between the engineering receiver and the destination. As he explains, "this semantic receiver subjects the message to a second decoding, the demand on this one being that it must match the statistical semantic characteristics of the message to the statistical *semantic* capacities of the totality of receivers, or of that subset of receivers which constitute the audience one wishes to affect" (Weaver 1949: 115). A second decoding: it is not enough for a purely probabilistic evaluation to take place, but a further statistical assay is required of the semantic fit of the message for an array of receivers or for that matter any definable audience in question.

Further, Weaver (1949:116) then introduced a new kind of noise—"semantic"—that he inserted in-between the information source and the transmitter: "the box previously labeled as simply 'noise' now being labeled 'engineering noise'. From this source is [*sic*] imposed into the signal the perturbations or distortions of meaning which are not intended by the source, but which inescapably affect the destination. And the problem of semantic decoding (and, indeed, semantic encoding) must take this semantic noise into account." In short, with Shannon and Weaver we never entirely get beyond meaning components and remain trapped in an intermediate phase where pure connective machinic potential is constrained by the vagaries of what Guattari (1977: 336) dubs "human 'understanding.'" Guattari (1977: 335) remarks on information theory that it "attempted to salvage something from the semiologies of signification in defining the significative redundancies as being in inverse proportion to the quantity of information." An increase in redundancy can help clean up errors, but it slows down processing time, decreasing the amount of uncertainty about messages. Whereas a decrease in redundancy enhances unexpectedness and hence the amount of uncertainty, so it contains more information.

What Guattari (1977: 336) would have us grasp is that "the remainders of a signifying process accumulate in the same manner as other strata of encoding. Lines of interpretance with their hierarchies of content, and lines of significance, with their controlled proliferation, become a kind of raw material for the construction of a-signifying sign machines." Enhancements of the components and their relations within the well-known point-to-point model, which became a specialty of sorts for Weaver, increase the representational redundancy of the model and limit its lines of proliferation, or at least slow them down by the process of assimilating semantics to technical issues. In other words, by the time Weaver finished with it, semantics mostly poses just another technical problem within an increasingly complicated signaletics; he matches the use of engineering as a clarification and specification of noise and reception with the insertion of new semantic features. So, the very factors that produce slowness also point forward toward intensive machinic productivity. In short, what's regrettable about the transmission model from a culture and communication perspective is promising from a machinic perspective. To add a proviso: by "inhabiting" such a model, a-signifying semiotic figures are not themselves prevented from undertaking phagocytic or parasitic activities (Guattari 2011a: 220). Moreover, the cyberneticization of the model by Shannon meant that the point-to-point sender–receiver could perhaps be automatically monitored and noisy messages "corrected." Shannon even imagined an observer ("auxiliary device") with the ability to parse channel capacity and micromanage the balance between rate, bandwidth, and signal power, in this way the ambiguities of semantics need to be translated into statistical trends in messaging and audience absorption levels. This so-called fuzzy remodeling (Seising 2009) was handcuffed by the addition of more and more components (in some cases [above] doubling components) whose machinic potential was not fully realized since the "subjective" observer is never fully automated except in select cybernetic systems where error-correction functions are integrated into the control system. The fuzzy line of escape ran straight into two constraining layers of personified components: senders, receivers, and the one "above" them both—the observer (who was not a robot just yet)—who directs the correction process. Such an "observer" may in fact introduce instabilities and delays into the system.

Tiziana Terranova (2004: 16) has astutely observed that the sender and receiver are allied, bringing to bear their different knowledges, in a single struggle against noise (distortions and diversions and uncertainties), and that the "minimal condition of communication" set out by Shannon and Weaver fundamentally displaces signs for the sake of signals, a kind of technified semanteme. This displacement is what I am identifying as an origin of sorts for a-signification. However, the "minimum" condition that corresponds to the predominance of signals should not be taken as a diminishment. Signals are in some respects scapegoated as the poor semiotician's stock-in-trade. What makes them less than signs proper is, as Umberto Eco (1976: 20 and 33) explains in *A Theory of Semiotics*, that they "can be computed quantitatively irrespective of their possible meaning." Signals thus on this view occupy "the lower threshold of semiotics." In my view, the lowly status of signals is unearned. Even if they slip under the semantic horizon, this gives them not only quantitative but machinically connective qualities worth exploring in depth. Signals may be revivified semiotically, rather than restored to full semiological status, through their a-signifying characteristics.

In the maieutics of Shannon and Weaver that gave birth to communication modeling, information theory is an obvious yet ambivalent point of departure for a theorization of the adventure of a-signification. However, it is also a quite useful one since it underlines some of the tensions, and it is to these constructive tensions and instructive entanglements that I want to turn in more detail as I lay bare the finer points of Guattari's development, conceptualization, and descriptive deployment of a-signifying semiotics.

Hybrid semiotics

In three books published originally in 1977 and 1978, in the two editions of *Molecular Revolution*, and in *The Machinic Unconscious* from 1979 (cited here in English translation 2011), Guattari elaborated a typology of semiotic systems framed in a Peirce–Hjelmslev hybrid conceptual vocabulary. Such a hybrid was already in use by Eco, among others, and constituted an alternative to semiological systems developed on the basis of Saussure's linguistics. Guattari (2011a: 18) himself rejects linguistic and semiological theories at the outset of *The Machinic Unconscious* because of their failure to consider "political, social, economic, and concrete technological domains which are in their common territory." A-signifying semiotics are defined relationally by Guattari against signifying semiologies, beyond which—outside the strata defined by content and expression planes and form–matter–substance relations—are a-semiotic encodings (genetic and other codes involving signaletic communication among viruses, cells, bacteria that are stripped of any projected "writing" onto them). Independent of language, then, a-semiotic codes consist of signals that "do not generate significations, do not engender stable systems of redundancy that someone might mistake as being identical to a representation" (Guattari 1978: 227). A-semiotic and a-signifying semiotics both deploy signals—but only the former is truly independent from representation and is manifested in a living being of some kind, taking on numerous substances (sonorous, chemical, gestural).

In spatial terms, then, a-signifying semiotics and signifying semiologies are located on the semiotic strata (of expression and content planes intersecting the hylomorphic features of matter–substance and form), and these strata are not isolated from one another. Like the Shannon–Weaver box diagram models mentioned earlier, Guattari's simple line diagram that I have discussed elsewhere (Genosko 2002; Guattari 1996a: 149), which is without a temporal dimension, but that can be added to indicate the processuality of destratification, indicates the skirting around of substance by form–matter functions. Indeed, what is instructive about Guattari's notion of semiotic strata is that they accumulate like "humus" in compost and break down over time (Guattari 1977: 332). This choice of language suggests there is something quasi-organic about a-signification or, put otherwise, that it is not only artifactual; better, that machines are not always mechanical and the machinic is irreducible to the mechanical. It is what the organic and inorganic examples share that interests Guattari.

Signifying semiologies concern well-formed substances (formed matter with a double planar articulation) situated on the stratified planes of expression and content, with the proviso that the transits among these strata are "centered on a single signifying substance" such as a sonorous substance of expression (Guattari 1996a: 150). Symbolic semiologies are a species of signifying semiologies and concern multiple substances of expression (gestures, ritual acts) that are neither completely translatable into linguistic terms, nor are they overcodable by any one substance of expression among them. This rule of nontranslatability and nonlinearity keeps at bay linguistic imperialism: "the semiological linearity of the structural signifier which imposes itself despotically over all other [non-linguistic] modes of semiotisation" (Guattari 1995: 49).

The distaste that Guattari (1978: 217) had for structuralist formalization may be appreciated in an example from a short talk he delivered at Douglas Hospital in Montreal (invited by psychiatrist Carlo Sterlin) in 1976 in which he maintained that the formulae of signifying systems obtained by structuralist analyses bearing upon language and the unconscious are analogically related to the digitization of an image by binary computer code: in both cases what is lost is the original "depth, warmth, and possibility of reorganisation." Of course, this is not quite up-to-date on the last point given the ease with which digital images may be manipulated, but Guattari's point is well taken in its context: it captures a reductionistic "technocratic vision of the world" that misses the "life" of language use just as diagnostic categories crush the ambiguities of real-life situations.

Guattari is never done with signifying semiologies; and in general one never really abandons them altogether. They are worked upon with, and sometimes against, often in a collision course. A-signifying semiotics puts signifying semiologies into play in some manner; in this way, a-signifying semiotics are not infected with semiological well-formedness, but it is something to which they may have recourse if communicating in the way that dominant significations require within a specific context. Children, immigrant populations, religious sects, and mentally challenged people can still express themselves without competence in signifying semiologies. But, Guattari (1996a: 151) boldly stated, a-signifying semiotics "can do without this [signifying] kind of crutch." The consequence of giving up the crutch, or not being able to acquire one permanently or in a moment of need, or even in outright rejecting one such as an immigrant enclave rejecting the mother tongue of its new home, is to come under the oppressive and sometimes invasive reach and greediness of certain reactionary signifying semiologies that will tell such a group what it means and what it wants (Guattari 1978: 229). Further, the consequences can be disastrous: "This relative untranslatability of diverse semiotic components is attributed either to a deficit, or a fixation at the pregenital stage, or a refusal of the law, or a cultural deficiency, or a combination of several of these elements" (Guattari 1996a: 152). In other words, the consequence is the application of a label based on some lack measured against some normative fullness that a given group has failed to achieve. Translatability into signifying semiologies becomes a yardstick for acceptable integration, the failure of which attracts a label like "potential terrorist" when an alleged cultural deficiency (i.e., rejection of "freedom") is combined with an attachment to another law (i.e., allegiance to an extremist sect).

Conversely, signifying semiologies are also capable of leaning on and "deriving their efficacity from the fact that they rely upon a certain a-signifying machine" (Guattari 1978: 236). That is, they may find the deterritorializing tendencies of a-signifying semiotics helpful in blurring the territories of the body (a cyborg's complication of the human and machine, including gender performances) or certain institutional spaces or eluding certain foregone conclusions, classifications, and categories. But in the very crossing between systems and generations of significations, new territories are breached and powers engaged, perhaps leading to the imposition of a more rigid definition, or conversely, to claims of incoherence. As the information model suggests, the intermediate position reveals that there is too much raw material to process, that the transformations of raw organic matter into humus have ceased, or that the further decomposition of humus has stalled as its stability has peaked and is no longer productive.

Guattari's (2011a:51) conceptual language sometimes includes examples from soil science, such as we find in *The Machinic Unconscious*: "a-signifying components develop to some extent on the manure of signifying components; they proliferate like microscopic parasites on modes of subjectification and conscientialization." Taken together with the automation of signifying semiologies by a-signifying semiotics, the growth of a-signification, like mushrooms on the excrement of signification, recommends the use of humification as a complementary term, and of the mixity of the semiotic processes that Guattari identifies.

Signification and power

The turn away from signs to signals does not imply the loss of a political perspective. Semiologies of signification are directly connected with power relations. Guattari is responding to both Foucault and Lacan in taking this position by detaching the subjugation of the subject by the signifier—the former insisting on this transformation, and the latter insisting on the letter's, that is, the signifier's constitution of the subject—and mapping escape routes from what he called the imposition of "dominant significations." Such escape routes can be extreme, he claimed, involving breaks with reality through mental anguish, drug and alcohol use. Guattari wrote (1978: 230–1): "signification is always the encounter between the formalization by a given social field of systems of values, systems of translations, rules of conduct, and of a machine of expressions which by itself has no meaning, as in a-signifying, and which automates behaviours, interpretations, and the responses required by the system." Escapes can be rerouted into power relations systematized in linguistic and other codes. Language serves capital, law, the state, official religion, etc. Guattari insists that the power relations at stake are controlled by the "ruling" class, in some instances the state. At the very least, national languages, whose structures are captured in the synchronic slices of structural linguistic models, crystallize power in syntagmas and syntaxes and grammars, the transgressions of which are subject to various kinds of corrective judgments, threats, and penalties: "The 'miracle' of capitalism is that it manages to direct language, as it is spoken, as it is taught, as it is adapted to its own evolution. Furthermore, this operation always appears to be self-evident" (Guattari 1996a: 143). For Guattari, meaning is not an effect of elements within linguistic structures; rather, it has a social origin in power relations to which signification is bound and without provoking a critical reflection on what is axiomatic.

Without meaning

The absence of a meaning dimension is less pertinent for Guattari than what is caught in the removal process: both image-based, that is, symbolically signifying, and mental dimensions or what Guattari refers to as redundancies of representation. A-signifying machines are not image-centered but diagrammatic (Guattari 1978: 234–5). This approach splits Peirce's inclusion of images and diagrams under the category of icons into two, because redundancies of machines can work without any representational dimension or other residual signifying effects. Guattari (2011a: 59) has us think of the "coefficient of deterritorialization" as a constant quantity that modifies variable sign machines, often by allowing them to act by duplication at places outside human perception. Hence, his penchant for soil examples. Guattari actively decenters enunciation from the human subject to machinic, nonhuman assemblages of proto-enunciation. Decentering human subjectivity for the sake of machinic proto-subjectifications is one of the broad theoretical goals of Guattari's philosophy. The field of a-signification becomes for Guattari (1995: 36) that of nonhuman enunciation in and among machinic systems: strictly speaking, "equations and plans which enunciate the machine and make it act in a diagrammatic capacity on technical and experimental apparatuses." This vast region includes everything from machine language "fetch and execute" routines, to system interoperability at different levels of exchange, or to multileveled cybernetic loops. These are scientifically formed by computer scientists and systems engineers. The convergence of a-signification and a-subjectification is achieved most clearly in the critique of anthropocentrism and the avoidance of anthropomorphism. Consider how Alexander Galloway (2012: 71) frames the discussion of computer code in software studies: against those who consider code to be illocutionary (performative language), Galloway situates code as firstly machinic and secondarily linguistic, which in Guattari's terms makes it more a-signifying than signifying, and that when machines instruct other machines no anthropomorphism is necessary to grasp the "functional nature of all software."

A-signifying semiotics, as I mentioned earlier, defines itself as signaletic. In this respect the non-necessity of semantic content in nonhuman communication is not construed as denying something to someone (i.e., to signal using animals, from birds to primates, and how these are redeployed across species, as opposed to the ability of immune cells to multiply protectively against an invading microbe) and does not entail some variant of behaviorism. However, this is a complex issue because Guattari's preference for ethological, not to mention microbial examples, is itself a deterritorializing move that is supposed to evacuate any residual "mind" from a-signification (of the sort that clings to senders, receivers, deceivers, and observers). This brings Guattari into the orbit of analytic philosophers of signaling evolution, such as Brian Skyrms (2010: 7 and 32), who, in claiming that signals transmit information but lack intrinsic meaning, retains plasticity of signaling without recourse to a mental element. This evacuation of philosophy of mind has a parallel in the evacuation of the individuated subject's fateful bond with the representational effects of the signifier, to which Guattari (1996a: 151) repeatedly refers in its Lacanian version: "A subjectivity individuates itself in the structures of this signifying machine, following the Lacanian formula that 'a signifier represents the subject for another signifier'. It is an ambiguous, duplicitous subjectivity."

By the time he wrote the last book published in his lifetime, *Chaosmosis*, Guattari had become much more focused on a-signifying info-networks and the devices used to engage with them, namely bank and debit cards. He had always shown an interest in examples drawn from the banking sector and economic institutions of capitalism, such as "the a-signifying grids [indexes, charts, data tables] of the Stock Exchange [*Bourse*]" (Guattari 1977: 237). Playing the market, or just as well being played by it, so to speak, was for him just a matter of "manipulating the a-signifying workings of the system" by any number of trading strategies, many of which are now recognized to be purely machinic involving the competitive advantage of high-speed computing—flash trading. A-signification is essentially informatic, not exclusively, or firstly, but presently and progressively so. Guattari consistently describes the assembling of partial or particle-sign components (parts that do not lack wholes and are fundamental and not grammatical particles) as a-subjective and machinic; in other words, as taking place without the mediation of subjectification at all and without forming a single semiotic substance since the idea is that substance is skirted around by form and matter. Guattari did not reduce his machines to technical devices, yet his repeated description of how a-signifying semiotics trigger processes within informatic networks highlights the interactions initiated with a plastic card bearing a magnetic stripe in activating automated access to a bank or credit account and engaging in an elaborate authorization process at a distance, which makes it clear that we are dealing with a complex, info-technological network. Guattari (1995: 49) clarifies that this has a direct purchase on material machinic processes like "a credit card number which triggers the operation of a bank auto-teller," algorithmically digit-checks the card number, opens access to resources through remote machine-to-machine communication and anonymization routines.

Part-signs

The key action of part-signs or particle-signs—signs that are partial (but not incomplete or completable), particle-like (whose existence is either hypothetical or simply minute), and destratifying—is triggering. This is Guattari's sense of molecularized signs: machinic superempowerment and diagrammatization. Having extricated himself from the Peircean trap of subsuming diagrams under icons (within Peirce's logic, diagrams are graphic representations—sketches, graphs, drawings, skeletons—in mathematics), Guattari wanted to then gain the positive implications of losing "aboutness" as a criterion, bringing diagrams into constructive coherence with a critique of representation. He split the image and diagram: the former belongs to symbolic semiologies and the latter to a-signifying semiotics. In shifting into a molecular-machinic modality of explication, Guattari highlights a tightly controlled repetition, whose deployment is extendable.

Interestingly, Guattari did not try to prove the existence of such part-signs beyond flagging a few examples here and there in his published works. Rather, for him the existence of such particles, like those of the hypotheticals of the subatomic world, did not need require "positive proof." He wrote: "Today one does not insist on providing a positive proof of the existence of a particle. It suffices that it can be made to function without any contradiction within the ensemble of a theoretical semiotic" (1977: 244). Experimental modeling does pose the problem of proof, but it is retroactive to the positing of existence by theoretical prediction or by machinic generation (i.e., by particle accelerators and colliders).

Particle-signs molecularize semiosis and are effectively blind to representation. They desubstantialize by emptying semiological and semiotic triangles (semiological triangles with thought at their summits, like the famous meaning of meaning pyramid of C. K. Ogden and I. A. Richards 1923: 11), both representamen–interpretant–object or form–substance–matter types. Diagrammatic particle-signs are dynamic and productive (capable of multiple articulations) but rigorously constrained—meaning is not essential in this activity, but specific codes, algorithms, materials, and standards are. Particle-signs work at the technomaterial level regardless of whether they signify something for someone or not: "A-signifying machines neither recognize subjects, persons, nor roles ..." (Guattari 1977: 237). Of course they can signify, since most users rely on some sort of mnemonic device to remember our passcodes; users probably need something of signifying semiologies given the proliferation of passcodes. As Guattari specifies, particle-signs do not in themselves "secrete significations"—whether these are "thoughts," "psychical" entities, or "mental" representations: "Signs 'work' things prior to representation. Signs (form) and things (matter) combine with one another independently of the subjective 'hold' that the agents of individuated enunciation (substance) claim to have over them" (Guattari 1977: 282). However, having incapacitated a disempowering representation and brought signs and things—the material and the semiotic—closer together, Guattari (1977: 344) then muses on sign-particle "dust" that emanates from the emptied triangles of meaning: "a thousand sharp points of deterritorialising particle-signs" pricking the spaces of abstract potentiality. This centrifugal force of particle-signs is described by Guattari as bearing a "quantum of absolute deterritorialisation" and is a "machinic superpower" that ordinary, individuated subjects cannot interrupt or tame, though they will try. Shannon's introduction of an "observer" who would feedforward corrections is a good example of what Guattari (1977: 345) envisaged as the erection of an "ideal point" upon which communication is concentrated and controlled. Yet dust-particulate is an important theoretical object. The sedimented city of metal dust, construction debris, chemical residues, airborne pollutants, and decomposing plastics has been joined by conjectures about smart dust particles enabling constant machine-to-machine communication by means of miniscule wireless sensors gathering environmental data (Gabrys 2010). Guattari was dreaming of a completely machinic self-sensing environment before its time.

Sign dust

Guattari's suggestive remarks about sign dust and the powder that spews from molecular machinic multiplicities, especially at large scales (i.e., the mineral and chemical deposits left on the skin after a day in most cities), may be concentrated in the way in which a-signifying semiotics eats into or empties out signifying semiologies. In the process, the particles generated are put into play in constructively alternative, nondominant ways; yet the dust may also blind and arrest movement, thereby impotentiating prospective escape routes; after all, the dust also contains debris from dominant semiologies. In *Cyclonopedia*, philosopher Reza Negarestani offers an analysis of the geo-mythic foundation of the Middle East as a "dust plateau" in the form of dust particles and fluxes and how they mix together with fundamental elements: earth, air, fire, assuming a paradoxical liquidity (water). Negarestani's original insight expresses a trajectory that is influenced by the combination of three elements with various kinds of wetnesses that are alternatives to water. He begins by pointing out that dust particles carry materials from many different territories and together creatively compose collective assemblages—what we call clouds, sometimes devils, other times metal and mineral traces in clouds; consider the fungal spores of San Joaquin Valley Fever and the manner of gathering matters from afar. Although dust is in this sense creative it is also rapidly dissipated; self-creation is linked to self-degeneration (Negarestani 2008: 91). Flux and reflux: dust returns to its nomadic existence. It reseeds that from which it emerged. This is a kind of creative nihilism: "it transcends itself back into the immanence out of which it came," as Brian Massumi puts it (2011: 22).

The first and most obvious moisturizer is oil. This has had many messy results in the desert wars of our lifetimes, not only in the Middle East, but in northern Alberta in the tar sands. But other "obscure wetnesses" may be involved and when these abominable particles settle in the earth, they contaminate it, until they dry out or are drawn out: "The carnal diagrams of flesh are imbued with dust soup" (Negarestani 2008: 94), like adulterated sausages and fraudulently doctored automobile transmissions, both augmented with sawdust. Pairing with what are called "questionable liquidities" revolutionizes particles, giving to them unusual propensities in the form of dry hails consisting of crystals carrying sand, like bloodred Saharan rains. The imperfect storm: the most deterritorialized and the most artificial signs in the Guattarian vocabulary are a-signifying, yet for all of the hope they bring in the form of escape from subjection into the exploratory openings provided by the digital universe, the molecular revolutionary potential they promote through the assemblage of particles is still only a tributary that flows back into a more dominant semiology. Backflow against the movement of tributaries is a form of signifying stagnation.

However, the point Negarestani (2008: 98) is getting to is that the ancient metaphysical diagram of earth-air-fire-water loses the "integrity and arrangement of their neighbouring and opposite elements" when modified by alien liquidities (tar sand) beyond wet and cold, the fourth element precipitating an "alien cosmogenesis" in combination with the "fluids" derived from dead seas and duststorms (various haboobs). The seeding of "otherwordly building processes" not only tangles the root elements but also intervenes in an established pattern by means of outsider impositions.

Alien liquidities are not so alien to ritual events such as weddings when the couple is showered in confetti, and in the circus, clown acts involving confetti hidden in what appear to be water pails are a source of humor and amusement, especially when tossed at an audience anticipating a soaking. Shredded and punched papers have a long history in public celebrations, and are described in liquid terms as showers, streams, and even flurries of "ticker-tape" falling on otherwise dry urban parade routes.

Inflected by Negarestani's hypothesis that a radical cosmogenesis occurs when alien liquidities combine with basic elements, Guattari believed in the necessity of a semiotics adapted to the most artificial environments perfused with sign fragments and particles, the understanding of which might cause a leap in semiotic functions of the kind suggested earlier in making dust particles communicate for environmental purposes or other pragmatic ends. *Smart dust*: instead of random specks of dust that stick to a digital camera's sensor, a tiny system of microelectromechanical devices, communicating wirelessly, clean the sensor of interfering materials. Not ambient particulate on the sensor, but dust as sensor. The collision course of micro-machinisms with dust announces the need for a future semiotics adapted to digitization, miniaturization, and other forces of the postmedia era. The control of such devices would not be, for Guattari, a forgone conclusion in favor of the military or industry, because the mobilization of molecular potentiality cannot be decided in advance.

Start and stop

Guattari's (1995: 49) a-signifying particle-signs "give out start and stop orders." It is easy to think of such particle-signs as the actual iron oxide particles on the tracks of the magnetic stripes of credit cards that are decoded—their polarities are immediately converted into binary digits when "swiped" by a reader with the appropriate software. As everyone knows, there is normally more to the operation than the gestural act; today, we are more likely to "tap" our contactless access cards on "terminals." Of course, Guattari's use of particles tells us that the signs of a-signifying semiotics are just as much virtual, "elementary" entities which are generated by machinic interactions like acceleration and mathematical prediction, and whose existence is verifiable theoretically. Indeed, particle-signs are the bearers of potentiality "beyond" the material fluxes and concrete machines that manifest them.

For the sake of generating concrete examples, my tendency is to proceed to the level of technomateriality: anyone who has received an error message during the process of inputting a PIN/password while undertaking a debit transaction or login operation understands the overt syntagmatic sensitivity of such signs (and in most cases the syntactical features—how many digits, upper- and lowercase sensitivity—of a password or PIN). Indeed, anyone who has ever had their card "eaten" by a machine or accidentally demagnetized knows the vicissitudes of a-signification—it may be just a jammed trigger, but it might also be a security countermeasure prompted by the card's use in a certain place, or for a certain purpose, inconsistent with an extrapolated pattern of usage. Moreover, when a card is, as one says, "all swiped out" by intense usage after a shopping spree, the kind of interaction between the oxide particles on its magnetic stripe and the card reader head that converts the encoding into binary digits goes awry because the magstripe is scratched or erased, thus introducing imbalance into the signal/noise ratio. Likewise, contactless smart cards conform to a number of international standards and protocols, operate within a fixed frequency in the case of radio frequency signal interfaces, and obey various wireless protocols, all the while transferring energy and data across a fixed amount of space. A-signifying part-signs do not slide; conversely, if they experience significant drift because of scratching, they cease working, or show signs of having been damaged, or hacked. Rebooting a nonfunctioning card can be quite homespun—rubbing on clothing, enveloping in paper or plastic, application of saliva—all of which are designed to manage a magnetic field.

Whether they are randomly generated or carefully selected on the basis of paradigmatic clusters of birthdates, children's ages, former addresses, initials, nicknames, etc., PINS/passwords, like the magstripe-reader encoding–decoding relation, can do without mental representations, which may of course exist, but they are not essential and no longer center signification. Passwords just allow one to pass through the gates.

There is a tendency in the information age for a-signifying semiotics to maximize its machinic force—to rapidly evolve, speed up, acquire greater mobility, miniaturize, and proliferate. In a-signifying semiotics, particle-signs work "flush" with the "real"; or more precisely, with material fluxes. Guattari does not, however, uncritically valorize flushness as directness. At the same level as and in parallel with are better ways to put it. Borrowing a notion from Peirce, even flushness does not always require physical contact, just an indexical contiguity that is not limited to proximity but has functional connectivity. For Peirce (1932), "an index is a sign which refers to the Object that it denotes by virtue of really being affected by that Object" (2.248) through a brute action that is *genuine* secondness. This underlines the networked nature of a-signification with select matters: it could be mycellium or silicon or laser light. In the example I have been presenting, the particles on a magstripe card can be demagnetized by exposure to more powerful magnets, which indicates the quality that sign and object share.

Diagrammatism, in Guattari's hands, blazes a trail beyond the human and individuated subject (of the statement) into the collective machinic dimension, escapees from the prison house of meaning: "We leave the terrain of signification," Guattari wrote (1977: 260), "for that of the plane of machinic consistency"; that is, the continuum of interactions on which any machine is reducible to an individual only arbitrarily, and where hierarchies like those of "reifying denotation and imaginary connotation are blurred" (1977: 259). On this plane are found events that are individuated and singular; it is devoid of formed substances; and it is neither anchored to subjects nor organized by imposed strata. With a-signifying semiotics one enters the plane of the post-human, "more and more artificial." Guattari didn't shed any "humanist tears" over those ill-adapted to such change, rejecting antimodern and antimachine recapitulations of humanism.

Semiotechnological software

In *Meaning in the Age of Social Media*, Ganaele Langlois establishes that Guattari's theorization of a-signifying semiotics has an important analytical contribution to make, especially with regard to the understanding of software as a technocultural agent. For Langlois, meaning is technocultural; its processes cannot be exclusively described as discourse, and its articulations involve nonhuman machines. In this sense, software is "a new kind of communicational actor" (2014: 52) that proto-enunciates. This semiotechnological object of software "makes sense for us" (qua users) and exploits in so doing both signifying semiologies and a-signifying semiotics. The production of meaning includes software that acts in the background, translates data into recognizable signs, and automatizes, tailors, and targets, but without thinking for it is "without any kind of human understanding, and representational and linguistic capacity" (2014: 70). This is a new Freudian "dreamwork": software does not think, but rather, it correlates, filters, and produces signs and signals; it may be useful to cite here Facebook's EdgeRank algorithms which establish affinities between user interactions and news feeds; that assign and calculate weights for activities, and prioritize time in terms of both recent and persistently active actions and relations as well as with a view to generating more and longer user activities online. Roy Bender and Neal Thomas (2014: 435) slyly describe this as "enticing engagement." An "edge" is an event on a social media site initiated by a user, and it is taken up as a particle-sign (clicking "like") that feeds the software ranking machine. What Langlois is most interested in is recommendation software on big corporate sites like Amazon.com. In telling users what they like, what they are searching for, and who they are, these proprietary algorithms implicate them in immaterial capitalist articulations of signifying and a-signifying processes. Human users are being actualized by nonhuman, machinic and a-signifying forces of organization, as consumers, but this is not the only subjective process available to them. Langlois writes: "a-signifying semiotics are not primarily concerned with meaning as the content of signification but with the adequation of a communicative ensemble with the real" (2014: 82). Flush with the real, a-signifying particle-signs affect users and they are affected by the choices of users, and the modeling of choice. Interpretation is eclipsed by the forces of repeat encounters.

Machinic liberation

Meaning may not be essential, but politics is. Put another way, even though Guattari moves beyond representation, the fact that representation is imbued with power relations doesn't mean that they are lost. For Guattari, all molecular phenomena display a politics in lieu of a signified. The particle-signs are no different in this respect, though on the face of it, the move to quantity and machinic interactions (automated triggers) belies it. Let's return to the magstripe. On the stripe, which is located in a certain position on the plastic card, there are several tracks. These are not neutral tracks upon which the magnetic particles are lined up. Rather, of the three tracks available, the first was developed for use by the airline industry, whereas the second is used by financial institutions. Each track's format was developed by and for specific interests. The cards meet a variety of international standards and function by means of specific algorithms. Recall the phrase quoted earlier: a-signifying machines may be used to "automate" the messages of the signifying semiologies that, in a capitalist system, begin stirring at a young age, especially around basic training in capitalist behaviors, namely credit, into which one is socialized. One could argue that the very agreements that permit these cards to work, namely standards, are a good example of what it means for any kind of sign to be flush with the world, but in virtue of international protocols and accreditations, quantified by International Organization for Standardization (ISO) designations (Busch 2011: 3).

A-signifying diagrammatic semiotics describes for Guattari (1977: 237):

> [...] the very texture of the capitalist world [...]. A-signifying machines recognize neither subjects, nor persons, nor roles, and not even delimited objects. That is precisely what confers upon them a kind of omnipotence; they pass through signifying systems within which individuated subjects find themselves lost and alienated. One never knows when or where capitalism ends.

A-signifying semiotics are perfectly adapted to the networked banking systems we use on a regular basis. Their diagrammaticity will mobilize the next extensions, not yet actualized, of cash networks and placements of automated transaction terminals, and new radio frequencies colonized by the next corporate players, and the coordinated triggers that open pathways through the expanding network. Guattari (1977: 345–6) explicitly turned to historical examples of banking systems (i.e., the Venice–Genoa–Pisa triangle in the Renaissance) in order to explain how the diagrammatic potential of this "liberation" of a-signifying machines was successively limited throughout the history of banking by serving the principles of oligarchy, debt, and centralization.

Today, the neologism "semiocapitalism" considers the latest phase of capitalism as a semiotic operator that emphasizes automated, machinic networks and interactions. Simply put, an info-commodity under semiocapitalism consists in a nonexclusive way of a-signifying particle-signs whose production and passage through digital networks contribute to the development of the machinic phylum, which is, for Guattari (1996a: 126), the creative historical force of "selection, elimination and generation of machines by machines."

The immaterial labor hypothesis picks up the Guattarian emphasis on the abstract machinic character of particle-signs, which is evident in Franco "Bifo" Berardi's (2007: 76) observations about the coagulation of capital flux in nonmaterialized info-commodities. I will return to this point in greater detail in Chapter 3.

Obtuse signs

The difference between a-signifying semiotics and signifying semiologies is established by a shared set of categories of classification; indeed, they occupy common strata. However, a-signifying particle-signs utilize signifying semiologies as tools for deterritorialization and for making novel connections between semiotic machines and material fluxes otherwise held apart within signifying semiologies (an individuated subject detached from the real and bewitched by representative images). Guattari's conceptual language extends to activity "triggers" (start, stop), sentinels on magnetic stripe cards, the devices that read them, and the networks that circulate the decoded data. However, as we have seen, the example of a-signifying mycellium feeding on nutrients—a rich semiological humus—before fruiting was equally relevant for Guattari.

The first question raised with regard to the relationship between these two kinds of semiosis is this: how do they relate to Barthes's model of semiological accumulation in the stacked and staggered systems of meaning? There is a second question. Does Barthes, with the concept of a signifier's obtuse meaning, achieve an insight into a-signification comparable to Guattari's? Barthes and Guattari are odd bedfellows: the former's linguistic semiology is at odds with the latter's mixed semiotic approach and development of a-signification. Yet Barthes, too, shared a putatively a-signifying semiotic sense in the form of obtuse meaning, even if he retained "meaning" as his point of reference.

In his *Mythologies*, Barthes describes how a first-order linguistic semiological system is built upon by a second-order mythological system by means of converting the unity of the first signifier and signified as sign (final) into a new signifier (first) for a second signified and unified sign. This is an operation of "construction," Barthes writes (1973: 123); it is the conception of final step as first new step. Myth capitalizes on semiological patterns and uses them as "raw materials" to erect a politics marked by a sly disavowal of ideology—what Barthes called "de-politicized speech."

By contrast, Guattari (1977: 245) assigned to a-signifying semiotics a disruptive and difficult micropolitical task of "eating into the semiology of the dominant order." In order to accomplish this task, it "will retain a certain partial use" (1977: 266) of signifying semiologies, which will always have a supportive but not central role to play in a-signifying proliferation. Guattari specifies: "the dregs of the signifier, figures of expression and prediagrammatic assemblages, are essential elements for the engineering of accelerators of particle-signs, the derritorialising power of which will be capable of smashing the strata of encoding" (1977: 336). Residues of signification (dregs, remainders) accumulate in the collapse of signifying redundancies, implosions of separated strata, collectivization of individuated human consciousnesses, and multiplication of double articulations (stalling this colonizing semiolinguistic machine). As destratification picks up speed and frees up more intensive processes, raw material for a-signifying semiotics is generated. This raw material, once assembled (self-organized and/or machined), is none other than the particle-signs that a-signifying machinic processes make use of; hence, my use of the term *sign dust*. Guattari writes: "Consequently, these territorial residues reorganize themselves into a-signifying particles; they will provide raw material for a-signifying semiotic machines beyond the reach of the impotentiating advances of reflexive consciousness" (1977: 341).

A-signifying semiotics are neither meta-codes nor models in the sense that myth is a metalanguage—Barthes's so-called second language. A meta-model for Guattari is critical of the model at which it points: it reads other modeling systems and through this activity a new subjectivity is assembled (Guattari 2013: 17). The model in this case is signifying semiology, which has a "limitless hegemonic claim" (Guattari 1977: 341) on signification. This very ambition is displayed by Barthes in his understanding of the "language-object" or linguistic sign in its globality (qua sign which "lends itself" to myth). However, myth is a colonizing force of language and it works by any number of tactics: parasitism, amplification, insinuation, conjuration... Ultimately, myth remains a signifying semiology. On this point, then, Barthes and Guattari diverge, despite the superficial structural similarity their thought displays in the categories of analysis and how they relate, though Guattari's is more diverse and shows greater resistance to semiological ambition.

Barthes's traits of the obtuse function at the level of the signifier. They are akin to Guattari's particles, but the latter are not creatures of the signifier. Rather, they are framed in terms of fluxes articulated by expression and relational (and reversible) content planes, and subsequently smoothed machinically (energized) as they are deterritorialized: "passive figures of expression are transformed into active particle-signs" (Guattari 2013: 88). In short, traits from signaletic fluxes are extracted into particle-signs and put into play in combination with fluxes of energy. Having broken from the linguistic signifier, Guattari's "'fundamental' entities" pass from extensive (space–time location and sensible traits) to intensive states (full of potential and multilocational) by means of the expression–content function and not the signifier–signified relation between psychical entities (sound-image and concept). Still, Barthes (1977: 61) is not content with a simple definition of obtuse meaning as a signifier without a signified because it cannot be named, is nonrepresentational, and eludes the language of criticism: "we do without language yet never cease to understand one another." Guattari takes much the same attitude: "It [a-signifying semiotics] can do without this kind of crutch (signifying language)" (1996a: 151). Barthes and Guattari are close to agreeing that the obtuse/a-significational is nonrepresentational, and that these figures are not easily absorbed into criticism, but with an important qualification. For Guattari, meta-modeling is a critical assay launched not from above but from among many models, taking aim at structural models across the disciplines.

Recapitulating his stacked systems, Barthes (1977: 52) proposes a three-tier system of meaning in his analysis of stills from Eisenstein's film *Ivan the Terrible*: the first is information or communication—"what I can learn from the setting, the costumes, the characters, their relations"; a second is a significational or stratified symbolic level consisting of various symbolisms—referential, diegetic, Eisenteinian, historical; and a third level of significance consists in signifying traits that do not yield a signified. Obtuse meaning is "excessive." Whereas obtuse meaning is, as Barthes remarks, "persistent and fleeting" (1977: 54), Guattari's particle-signs are reticent about signifying anything to anyone; however, they, too, are traits or parts. Barthes, as well, enlists "indifferen[ce] to the story" (1977: 61) as a feature of obtuse meaning. And there is more to note. Barthes insists on the "im-pertinence of the signifier" as a robust feature of indifference to the obvious meaning of a story. He describes it as a "de-naturing" effect, a "distancing" from the referent by means of intense sounds and colors without "natural" reference points. These remain "depleted."

On this medium Barthes and Guattari converge: film is a prime site for a-signifying semiosis. In discussing *Badlands* (dir. T. Malick, 1973), Guattari (1977: 206ff) insists on what most critics missed: the agonizing blue of the enormous skies of the location; the *amour fou* of the young couple; the a-signifying connections that go nowhere (the father's murder—retrieval of a toaster from the scene and its relocation to an encampment without electricity). Rather than drawing upon Barthes, Guattari draws upon Christian Metz for his explanation of the a-signifying fabrics (sonorous and visual) of cinema that resist signifying semiologies. This is what Barthes (1977: 65) called the filmic as such, irreducible to the film: "The filmic [...] lies precisely [...] in that region where articulated language is no longer more than approximative and where another language begins (whose science, therefore, cannot be linguistics, soon discarded like a booster rocket)." Guattari finds in Metz the importance of film's images that are "matters of content" which remain undefined, and "matters of expression" that are unfixed. But Barthes never really discards linguistics, building a vast semiolinguistic universe of interpretation. At the same time, Barthes isolated certain outer limits beyond the strap-on rocket stage in his own practice while retaining—and this is an obvious point—obtuse meaning (extra-structural) beyond linguistically articulable and structuralizable meaning-effects. He reached the margins by focusing, self-consciously and paradoxically, on film stills, in order to grasp the processes of moving pictures. Obtuse meaning makes meaning into an irritant, and this is its a-signifying force.

The crossing points between Barthes's obtuse meaning and Guattari's a-signifying semiotics are sufficiently dense as to warrant close consideration. Just as I showed with regard to the remodelings of Shannon and Weaver, the invention of a-signification with Barthes in the equation remains stuck in an intermediary stage of advances and rearguard actions in the name of meaning. The eclipse of meaning, as Guattari reminds us, is never accomplished once and for all. It is not so much a temporary obscuration as a counter-hegemonic destabilization and decentering of signification that opens up hitherto closed routes of escape, but is not itself immune to cycles of liberation and recapture.

Neoliberalism as an a-signifying regime

Starting from the destructiveness of entrepreneurial subjectivation, Maurizio Lazzarato, in *Signs and Machines*, considers neoliberalism to offer "no new production of subjectivity" (2014: 8). The entrepreneurial model, conversely, impoverishes social life and promotes depression, precarity, poverty, and debt. This is a repressive and regressive or "negative subjectivation." While Guattari (2013:13) looked to Japan during the 1980s for an example of a contemporary cocktail of subjectivity combining archaic elements and hypermodern, hi-tech currents, Lazzarato considers the post-bubble economy of Japan to be a "crisis in the government of behaviour" wherein "economics and subjectivity go hand-in-hand" (2014: 11). Subjectivity is produced, as Guattari and Deleuze argued, by two means: social subjection and machinic enslavement. Lazzarato's original interpretation of the contemporary significance of these apparatuses reveals their fundamental tension in the era of neoliberalism: the first assigns identities, and the second dismantles them. A most ambiguous designation like "content-provider" is both a social assignation within web development, but it is a name for discontented self-management on the margins of the new economy. Subjectivity is produced at the intersection of subjection—individualities like citizen, Canadian, worker—and enslavement diverges from representational semiosis and breaks it into parts. Enslavement controls and manages parts (cogs and components) of an integrated system. In such a system, labor itself is divorced from productivity and surplus value is produced without employment, but by watching television, clicking "like" buttons, blogging and vlogging consumer products, swiping loyalty cards. Lazzarato isolates the lack of proportionality between labor and production, workplaces without boundaries, and the never-ending workday. What we are presented with is the new reality of capitalist exploitation of the difference between subjection and enslavement: the measurability of a social function gives way to the "never assignable nor quantifiable" (2014: 45) dimension of machinic enslavement in which productive activities are carried out beyond employment relationships. Indeed, the entrepreneurial management of the self no longer leads to employment but to taking on the burdens, that is, risks and costs, of what neither the state nor business are any longer willing to bear (2014: 53). Later in this book, in Chapter 3, I will take up this argument with regard to semiocapitalism.

In the meantime, Lazzarato explains Guattari's major contributions to critical semiotic theory by emphasizing the insight that invariance, stability, and automation are the cornerstones of the establishment of dominant significations that shape a standard kind of subject. The mediation of a representational element between signifier and referent is a disempowering split that emphasizes mediation (consciousness of the individual and the psychic status of both signifier and signified) over a direct purchase on the real, creating a gulf between semiosis and the real; henceforth, referentiality simply generates quandaries. Lazzarato underlines the consequences of the impotence of signifying semiology by contrasting a more open kind of expressivity of the kind found in so-called primitive societies where symbolic semiotics remain polyvalent and heterogeneous at the levels of multiplicity of referents, indicating multiple states of reality but also the preponderance of the group/community over the individual enunciator, the latter including an alignment of the human with nonhuman actors. Further recourse to anthropological material will be linked in a number of the following chapters, particularly focused on Jean Baudrillard (Chapter 2) and Jose Gil and Claude Lévi-Strauss (Chapter 6), to the critique of semiosis.

Guattari's main contribution to sign theory is, for Lazzarato, a-signifying semiotics, but understood machinically. As Lazzarato points out, a-signifying semiotics include: "stock listings, currencies, corporate accounting, national budgets, computer languages, mathematics, scientific functions and equations…and [examples from] music, art" (2014: 80). A-signifying semiotic machines involve the enslavement of human beings through their combination with certain nonhuman components in ways that position them as parts of integrated functions. The ontological implications of this position are profound: "In machinic enslavement, the ontological barrier between subject and object established by social subjection is continually blurred not because of language but because of a-signifying semiotics" (2014: 83). Such semiosis permits machinic assemblages to enunciate nonhumanly and these enunciations (often qualified by Guattari as "quasi- " or "proto-") are creative and productive, they make things happen, produce changes by directly engaging real, material flows. Lazzarato, too, gives the example of the magnetic stripe card, and reminds us of Guattari's example of what his parking permit allowed him to do: to pass from outside to inside through a gate. The emphasis is placed on operationality as opposed to representationality. Machines communicate with one another, with ourselves, and we communicate with them, but not always and not necessarily through language and our five (or more) senses. In my home I have sophisticated devices that detect trace amounts of radon gas and carbon monoxide. These devices sense what I cannot and alert me to their presence by issuing warnings. In *The Making of Indebted Man*, Lazzarato (2011: 148) uses the automated bank teller and bank card to move from an explication of debt as subjugating to machinic enslavement as passifying: no "individual" is at work before the ATM, just a "dividual" cog in a sociotechnical apparatus.

Lazzarato underlines the role of "partial discursivity" in machinic enunciations, raising in passing an important issue about Guattari's understanding of discourse. Guattari separates discursivity from pathic apprehension of the world. Discourse is for him characterized by locatability in terms of space–time–energy coordinates, sequences, linear chains, linkages such as clusters. These are exo-references. Pathic apprehension is endo-referenced, noninterpretive, affective. Yet discursivity inhabits pathic being, and supports the very content that it tends to evacuate and bracket in search of other extrinsic evidence and markers of order and differentiation. Guattari wants to discover how the discursive and nondiscursive realms interact. Just as he notes with regard to how a-signifying semiotics are never done with semiologies of signification, likewise pathic apprehension is never finished with discursivity, yet it is not fixed by it, in the same way that a-signification is not fixed by meaning. Why? Because nondiscursive experience, in all of its immersive, immanent, and chaotic heterogeneity, needs discursivity to discern its becomings, slow down and express the vitality of its de-differentiated world (its fecund moments). The dynamic interface between discourse and pathic life, when one is plunged into chaos, and resurfaces to find some way to understand the intensity of such experience, invariably returns to discourse. Discourse is not the enemy! (Guattari 2011c: 36) It inhabits pathic being in small amounts. It cannot be done away with, only diverted from its overt ends—its strict denotational, broader significational perspectives and bureaucratic applications (see Genosko 2015). Diverting from denotational and connotational entrapment in semiological strata, as Barthes proposed in the staggered systems of *The Fashion System* (1990), raises the question: divert away from, but toward what, exactly? Divert and find support: these are closely related ways that pathic intensity self-enriches through discursivity, despite discourse's tendency to bracket pathic experience (restricting its creativity in the service of scientific models). The goal of this diversion is a refrain, a pathic temporalization of being that gives consistency to sideways tangents and escape routes. Anne Querrien (2011), who worked with Guattari at the Centre for Institutional Study, Research and Training (CERFI), provides an example in terms of writing as a bureaucratic form of expression that shifts to an adjacent position with regard to its overt purpose or service role. In the process, its tangency is enhanced and its escape into other genres within or beyond sense is supported with consistency, achieving some measure of autonomy and generating surprising affects. Guattari poses a paradox: the foundation of subjectivation is pathic, and the pathic contains "small supplies" of discourse, which as they expand tend to squeeze out the pathic by recourse not only to rationalistic prerogatives but badly botched narratives like consumerism and celebrity culture; yet, discourse also rests on the very thing it attempts to evacuate, namely, nondiscursivity. And, conversely, the nondiscursive requires discursivity (i.e., some sense of succession or order) along the way to making discernible its heterogeneity.

The human-machine assemblage is at once the site for the capitalist exploitation of a-signifying semiosis with its engineered subjectivities and the promise of new collective formations that can act against capitalist subjectivations. For Lazzarato,

> A-signifying semiotics and technical, scientific, artistic, and revolutionary processes of deterritorialization constitute the propitious conditions for doing away with the humanist, familialist, and personological modes of representation, the nationalistic, racist, and classist modes of subjectivation, according to which capital is deterritorialized and in which individuated subjects become alienated. (2014: 94)

Both of Guattari's favorite examples—one we have already discussed, the informatic network and the interface of the magstripe card, which expands to include all similar actions and machine-based interactions throughout our lives, and the other, driving a car—are primary examples of capitalistic machines. Lazzarato spends some time investigating Guattari's description of driving an automobile as machinic enslavement that involves partial-enunciations—nonreflexive human and advice-dispensing vehicles (voice-activated infotainment as an "option," seats with "memory," quasi-autonomous, self-parking functions) that may of course be scrambled when a problem arises (requiring a specific stratum of workers to clean up or repair the damage), but much of the time mesh together in quite routine ways. Drivers are not users, that is, they are not subjected to their machines; rather, they are enslaved by them. Lazzarato pushes forward Guattari's example and brings it into the present, extracting it from the position it occupied in Deleuze and Guattari's *A Thousand Plateaus* as an example of a subjugating technical machine birthed by the modern state (Fordist mass production), whereas machinic enslavement was much more ancient as in the megamachines of Egyptian slave labor described so elegantly by Mumford, a favorite of Guattari's, and it has returned with a vengeance in the subtle networked codings, the "systemic entanglements" of the computerized planet (including wireless vehicular navigation, safety and security systems; Deleuze and Guattari 1987: 457; and Holmes 2009).

Lazzarato (2014: 90) cites the "multiplicity of modes of subjectivation" available in the machinic assemblage in order to interrogate the range and subtlety of capitalist exploitation, but without addressing concrete tactics on the vast machinic phylum of contemporary life; yet, he also seizes upon Guattari's banking examples to construct the figure of the stock trader as a human-machine using the displacements of the individual and human: the trader is only a "terminal" for rational and affective sign-flows, that is, for both signifying and a-signifying semiotics in which the representational (impotent, illusory signs) and transformational (power signs like diagrams) cross and give consistency to this kind of subjectivation (2014: 99–100).

Large-scale machinic enslavement

Guattari's fascination with Mumford's idea of "megamachines" poses a question of scale. Guattari uses it to describe large-scale entities like cities: "Cities have become giant machines, 'megamachines' ... producers of individual and collective processes of subjectivation by means of collective apparatuses (education, health, social control, culture) and mass media" (Guattari 2015: 105). Indeed, for Guattari (1986: 460) "the main function of the city as a node is the production of subjectivity." This "city" is nonspecific, an abstract entity. Like Mumford, Guattari found that the megamachine persisted into the twentieth century. But Mumford's tracing of the megamachinic trajectory along military-labor lines resulted in the big science that produced the atomic bomb. Mid-twentieth-century America was the site of new megamachinic nuclear and Cold War bureaucracy, governmental institutions built on secrecy, dehumanization through extermination fantasies, a well-funded priesthood of technoscientists, and a new divine king: the computer. In *The Myth of the Machine: The Pentagon of Power* (1970: 273ff), Mumford arrives at a delirious analogy: the "Omni-Computer" is our techno-God, all-seeing like the Sun God Re, monopolizer of power (information and knowledge), pervasive, invasive, controlling and destroying the human soul. This is a vision of a processed world in which there is "a total destruction of autonomy" (1970: 275) that gives rise to an ideal type: "Organization Man," both producer and product of the new megamachine age, a "virtual automaton" and "servo-mechanism" of the info-machine.

Mumford would not have understood Guattari's positive assessments of what he called the machinic enslavement of molecular components of preindividuals; he would not have accepted Guattari's post-human, decentered, desubjectified hybridity of humans-machines. Yet Mumford alerts his readers to a historical example of machinic enslavement where the machines are humans, and the new megamachine is, as Lazzarato explained, "the mode of control and regulation ('government') of a technical or social machine such as a factory, business or communications system. It replaces the 'human slavery' of ancient imperial systems (Egyptian, Chinese, etc.) and is thus a mode of command, regulation, and government 'assisted' by technology and, as such, represents a feature specific to capitalism" (Lazzarato 2014: 25–6).

Would Mumford (1970: 317) have labeled Guattari's invocation of a machinic phylum as "technocratic mysticism" in the spirit of McLuhan or Teilhard de Chardin? Guattari (1996a: 267–8) identified a vast mechanosphere of machinic evolutionary trajectories wrapped around the biosphere of the planet: "... not as a constraining yoke of an exterior armor, but as an abstract, machinic efflorescence, exploring the future of humanity." Many different types of machines assemble, some taking the lead, others falling in behind, on multiple axes (synchronic and diachronic) to constitute an object or series of objects; think of a Concorde jet (desire to economically fly a supersonic passenger jet) and the Apollo space program (J.F. Kennedy's political machines). These emerge from the phylum, as Janell Watson (2009: 161) correctly observes in her *Guattari's Diagrammatic Thought*, and are defined diachronically as the evolutionary descent of machinic species. What comes after the Concorde? The answer will be more than technical. Can humans land on Mars? Can humans live on Mars? Again, the economic, political, material, and semiotic components of an assemblage of machinic components are all at play in this journey. Guattari clarifies: "a machine is something that situates itself at the limit of a series of anterior machines and which throws out the evolutionary phylum for machines to come" (1996a: 126). Further, "a machine rises to the surface of the present like the completion of a past lineage" (1996a: 267), with the added observation that its dominance is provisional. The diachronic lineage ascribed to the machine is not a history written by the victors in the form of a golden legend of how they achieved their greatness! Rather, it displays a nonlinear and mutational rhizomic character, rather than a Darwinian tree of life.

Guattari is not describing a noosphere in the manner of Chardin as machines are not reducible to a conscious mind wrapped around the planet, which Mumford also dismissed as an "etherialized version of the megamachine" (1970: 314). Yet the emergence of the noosphere is an evolutionary process according to Chardin, not unlike Guattari's machinic evolutionism. The latter's emphasis is on an immanent multiplex machinism in the midst of which a "new alliance with machines" will need to be made on a planetary scale (1996a: 267). For Guattari, "it will be possible to build a two-way bridge between human beings and machines" (1996a: 95) and on this basis forge new forms of machinic subjectivity. Guattari's (1996a: 96) explanation of how subjectivity "enters the machine" suggestively draws on "entering a religious order"; it has an initiatic dimension. Machines constitute "collective apparatuses of subjectification" that, under the conditions of planetary computerization, generate digital-processual subjects from the earliest ages when children are given their first mobile phone or sat before a television set. This process, in Mumford's estimation, is a disaster for the planet because both docility and fidelity are required of human beings with regard to their machines, a position that he pins on Marshall McLuhan (Mumford 1970: 339). Guattari was not naïve about the machinic isolations, addictions, infantilizations—even "bribes" that would occur (Mumford's version of this concerns how automation creates dependency and keeps children under "remote control").

As noted earlier, there is a diachronic dimension to the mechanosphere (the machine that replaces another and will itself be replaced in turn) and a synchronic dimension that is expressed in a media theoretical sense of "network" and its inseparable machinic nodes. Guattari insists on considering how all machines fit together rather than cutting them out and analyzing them on their own. Occasionally, and unpredictably, a really revolutionary cut tears the fabric of the machinic continuum consisting of all the different kinds of machines (i.e., ancient revolts against Ptolemaic rule; Lenin's revolutionary "cut" as the bipolarization of the proletariat and bourgeoisie). For Guattari, nothing is ever lost once it appears, and the phylum is irreversible.

In *Schizoanalytic Cartographies*, Guattari writes: "At the limit there is just one machine on the horizon" (2013: 74). Continuing, he specifically rejects Mumford: "Not as the science-fiction of yesteryear imagined, in the form of a tyrannical mega-machine, but as a powdery molecular machinic multiplicity." Guattari may have rejected Mumford's characterization of megamachines as tyrannical, both ancient and modern, but he still utilized the ancient version of it as an early example, maybe the first, of large-scale machinic enslavement (Deleuze and Guattari 1987: 428). To be a "piece of a machine," to be "enslaved by it," is machinic enslavement: "We believe that Lewis Mumford is right in designating the archaic empires megamachines" (1987: 457). It would be incorrect, however, to simply designate modern technical machines as enslaving in the sense that Mumford suggested; rather, Deleuze and Guattari state, modern technical machines engage in the social subjection of workers or users of external/exterior objects. Of course, social subjection defines the institution of ancient slavery as well, but since human beings are components of the human-machine, they are enslaved by the megamachine. However, modern states also have the capacity to reinvent machinic enslavement, but this "in no way represents a return to the imperial machine since we are now in the immanence of an axiomatic, and not under the transcendence of a formal Unity" (1987: 458). Given the absence of such transcendent unity, thus rejecting both ancient religico-myth and Mumford's computer-God, contemporary, automated, cybernetic communication involves the coexistence of both subjection and enslavement, especially in television viewing: "one is subjected to TV insofar as one uses and consumes it, in the very particular situation of a subject of the statement that more or less mistakes itself for a subject of enunciation ('you, dear television viewers, who make TV what it is ...') ... But one is enslaved by TV as a human machine insofar as the television viewers are no longer consumers or users, nor even subjects who supposedly 'make' it, but intrinsic component pieces, 'input' and 'output,' feedback or recurrences that are no longer connected to the machine in such a way as to produce or use it" (Deleuze and Guattari 1987: 458).

Postmedia ecology

The inseparability of the processes of subjectivation and machines is Guattari's starting point for the exploration of how to encourage the production of incomparable and polyphonous modalities of a subjectivity that would be collective and not fundamentally a person, yet self-positing and constellated with other subjectivations and existential components. This is a selected precariousness in which assembly is required.

In a machinic ecology of relations defined by subjection and enslavement, media plays an important role for Guattari. Alienation from the mass media is a given for Guattari; his strategy is to gain access to newer media (databases, video, audio, all modes of interactivity) that dissipate passivity and release the hostage-subject of media exploitation, ensnared by the Western metanarrative of terror and security and the cynical and ethically indifferent subjectivations of consumer capital. New media can assist in the formation of unique alliances, even around basic services that offer a greater degree of control over programming and scheduling, or ambivalent practices like channel surfing (Guattari 1990). Shifting into a postmedia era is an antidote for the mass-mediatized status quo, but for all the hope that the aesthetic reappropriation of media production by video artists using television differently, computer-aided design, and hypertext held for Guattari, resingularization remained uncertain. All of the then current promises of the late 1980s and early 1990s media that Guattari wrote about did help to incubate new modalities of subjectivity and instantiate a new social ecology among producers and consumers realized in some ways by tactical media experiments in radio and television and later by crowdsourcing, citizen journalism, Web activism, and peer-to-peer file sharing. Guattari (1996a: 98) remained sober about the real prospects for change in the postmedia era. He was inspired by concrete poetry, early Rap music, spoken word performances in general since they augur experimental machine-voice concatenations. The growing power of machines to enunciate a-subjectively was an acceptable and desirable consequence for Guattari of the human-machine hybrids he imagined in the post-media era. Machinic enslavement is at times hypnotizing, and truly enslaving when, for instance, social media users volunteer personal information and preferences with the outcome that their choices can be premediated for them in terms of targeted advertising, personalized recommendations, and similar pseudo-singularizations. In terms of strategic necessity, Lazzarato states: "We must seize the opportunity of desubjectivation opened by machinic enslavement so as not to fall back on the mythical-conceptual narratives of producers, workers, and employees. This is one of our most urgent tasks if we are to invent new political subjectivations" (2014: 122). Without the subject's attachment to specific roles and identities from which to become alienated, enslavement permits an escape from social subjection, but with the requirement of recognizing that traditional conceptions of the human subject must be decomposed with the accompanying loss of anthropocentric perception through machinic—not necessarily mechanical—but animal or vegetable becomings. But the shadow of social subjection is long and humanist suspicions of any kind of enslavement are difficult to dissipate.

Machinic futures

For Guattari, the "liberation" of an a-signifying semiotic machine often runs straight into a species of capitalism that brutally exploits it. This course can simultaneously remove us from and bind us to specific models. In the late 1970s, Guattari developed a distinction between signifying semiologies and a-signifying semiotics in a manner that "remained very schematic", in other words, insufficiently mixed: "a signifying semiology is always haunted by a sign machine and, conversely, an a-signifying sign machine is always in the process of being recuperated by a signifying semiology" (1977: 346). Of course, he identifies polarities (political) and specifies the apparatuses of capture in double articulation, how a language should be spoken, and the overcoding and axiomatization of intensive means of escape. The creative freedoms of a machinic diagram may be stratified and rendered impotent, yet the repeated and creative assertion of such freedoms is in no way precluded. Guattari asserted that there was no "dialectical synthesis," no *Aufhebung* (1977: 356). Because a-signifying semiotics connects with "traits"—the particle-signs that are un- or only partially formed both semiologically and physically—in which a distinction between expression and content is not yet definitively operative (Deleuze and Guattari 1987: 347), it may experiment with how particles connect and enunciate beyond and before the human, as it were, before becoming tangled in the binding mesh of representation, repression, organizing and transformative subjectifications of pronominal voice (the splitting and de-diagrammatizing of "it" by the "I-ego") (1977: 347). Guattari wanted to show how a-signification cuts across the strata, swerving around substance, from which it makes its escape by forging machinic connections. This diagram has many iterations; for instance, in *The Machinic Unconscious* and later in *Schizoanalytic Cartographies*, the swerve is the main focus and the background is absent. Recall how Shannon and Weaver's additions to the transmission model—the qualified relabeling of existing, and the introduction of new, semantic components—compromised the machinic logic of the original, generating what Guattari would describe as an involutive effect: the implosiveness of a modelization that attempts to deepen and justify the irrelevance of meaning for transmission by absorbing "meaning" components into its technical desiderata. We also saw Weaver's slipping of residual "minds" (Baecker 2013: 87) into machinic communication, not to mention Shannon's surveillant apparatus: an all-seeing "observer" and different kinds of error code control management.

Yet this gerrymandering nevertheless spreads the elementary "dust" of particle-signs, which stick to the components and have the power to erode them, to disaggregate assemblages by decentering mental representation. As Guattari put it, "in diagrammatism, substantial semantic or signifying residues of the object [denoted or represented] and of the means of expression are always superfluous. Semanticism or significance are only tolerated in a provisional way, and the expectation is always that they will be reduced at the next stage of technical and scientific progress" (1977: 336). Such an expectation is not always met and is subject to skirmishes, rearguard actions, and partial advances.

In the informational space, to use Terranova's expression, of a-signifying semiotics, probability displaces signification and this pushes representation into crisis, in the way that, for instance, computational photography abandons perspective and realism in making images through redefinition and refinement in which not only linguistic, but visual representation is recast algorithmically so that images may be understood relationally and probabilistically (i.e., porn-detection algorithms) (Steyerl 2014).

Taken together, Guattari and Deleuze's remarks on control societies contribute to a critical understanding of what it means to enter a world where passwords—access and denial—form a high stakes technopolitics which cypherpunks phrase in a somewhat outmoded language of individual versus mass surveillance—the interception and storage of telecommunications data—but which, nevertheless, awaits the creation of the analytic tools that can trigger specific actions to exploit the situation (Assange et alia 2012: 37–40). Can a-signifying semiosis vouchsafe a revolutionary role in popularizing cryptography? Following Guattari, the sharp-edged particle-signs radiated in the process of emptying the semiological strata, and emitted from the black holes of impotence and disempowerment, remain liberatory in their promise of creative transformation toward the autonomy and anonymity of so-called personal information. Perhaps now we are all cypherpunks in training, and our politics is a struggle over the control of the alternatives with "a stab at the fabric of possibility" (Terranova 2004: 27) presented by a-signifying semiotics.

The originality of Guattari's philosophy of the machine is that it departs radically from the tradition, stretching at least from Kant to Heidegger, of conceptualizing machines "privatively": the devalorizations include an inability to act themselves and lack of a principle of unification, and, despite these, they pose threats of various kinds to humankind (Welchman 1997: 211–12). Guattari's insight was to imbue machines with a special semiotic power by borrowing and modifying Peirce's iconic diagrams and making them powerful, dynamic, and operational: generating reciprocal relations between semiotic and material flows (Watson 2009: 50). In addition, he adds that a-signifying semiotic signs are akin to signals, emphasizing their "non-human" features; indeed, Guattari's analysis moves through analogies grounded in organic media, namely, mycorrhizal nets of fungi, to digital networks of the information age. Displacing the semiotic centrality of language and signifying semiologies, but not absolutely, Guattari embeds a-signifying semiotics in a typology that reveals a nuanced and tensile set of relations. In shifting enunciation away from human subjectivity, and excising the requirement of representation (thus minimizing the gap between representation and reality, subject and object), the human-machine relationship is revalorized in a non-threatening way, creatively regaining cybernetic enslavement and the problematic of automation. Unbuilding the bodies of signs by positing the power of partial-signs, Guattari pursues a hypothetical trajectory of increasing abstraction, but one that may also be grounded in technomaterials such as metal particles on recording devices. Meaning remains an "irritant" for Guattari, and similarly for Barthes in theorizing the obtuse sign. What makes meaning irritating is that its encodings are aligned with dominant values and self-evident truths of capital, mass media, and power. One of the key tasks of a critical semiotics based on mixed signs is to investigate the collusion of signification with capitalism.

Given that Guattari relies on the concept of the megamachine in order to describe large machinic assemblages at the scale of urban agglomerations, he has a specific goal in mind: to discover the ways that cities produce subjectivities. He writes,

> The material infrastructure, communications and services of cities cannot be separated from functions that may be described as existential. Megamachines model sensibility, intelligence, inter-relational styles, and even unconscious phantasms. Hence, the importance of bringing about a transdisciplinary collaboration between urbanists, architects, and all the other disciplines of the social, human, and ecological sciences. (2015: 105)

Guattari wanted to influence the production of subjectivities within contemporary cityscapes and he took a Braudelian route, of sorts, to do so. He extended Braudel's invocation of the multidisciplinary requirement expressed in *A History of Civilizations* (1994: 9) that all the social sciences must be marshaled together in the study of civilizations, adopting his own term "transdisciplinary" to indicate it would be more than a multiplicity of expertise that would be called upon. The creation of vibrant microspaces of experimentation, in which Guattari engaged throughout every phase of his career, would be brought to bear upon the urban ecological predicament in global context (i.e., nontechnocratic solutions to industrial pollution; invention of nonpolluting transportation systems) in order to give direction to the urban machine's collective self-elaboration toward an ecological consciousness engaged in ethically responsible negotiations of collective actions and large-scale engagements. Since subjectivity is intimately imbricated in mutually dependent bio- and mechanospheres, Guattari called in his book *The Three Ecologies* (2008) for three ecosophical registers (mental, social, environmental) that would aim at an ethico-political linkage of the micro- and macro levels, building a critique of technocratic solutions and highlighting the role of artists in fostering emancipatory eco-praxes. Art and ecology are closely linked in the production of subjectivity in a way that would assist in extracting potential for existential change and assisting in the development of new processes that are more complex, sustaining, and enriching. A new Guattarian megamachine requires the creation of collective assemblages of enunciation based on eco-political and aesthetic principles that promote resingularizing subjectifications. The overlapping registers entail that any environmental efforts aimed at global warming or species extinction must be accompanied by a change in collective consciousness and social relations; otherwise, fighting pollution is just "remedial" (2015: 106). For Guattari, it is the city that is at the center of the tri-ecosophical problematic because "the urban phenomenon has changed nature. It is no longer one problem among many. It is problem number one: the problem sitting at the crossroads of economic, social, cultural and ideological stakes. The city forges the destiny of humanity ..." (2015: 107).

If there is only one machine in the future, it is a planetary city whose "diverse components are scattered over every surface of a multipolar urban rhizome encircling the planet" (2015: 101). Drawing on the existing north/south divide and rich/poor distinctions, Guattari moves in the direction of a subjectivity that has escaped from capitalist valorizations employing these and other similar binaries and answers to other values based on solidarity, emancipation of oppressed groups, social reinsertion of the aged, etc. He rejects the universalist presumptions of architectural modernism in the manner of Le Corbusier for a more nonspecialist attention to the qualities of the subjectivities they aspire to produce through particularities of sites and inhabitants. These qualities themselves carry proto- and partial-subjective functions and cannot be abandoned to the whims of gentrification, market forces, consensual tastes, and so-called democratic processes of decision-making easily manipulated by financial interests: "This partial subjectivation, in a sense, will have a tendency to cling to the past, to some cultural influences and reassuring redundancies, but, in another sense, it will remain attentive to elements of surprise and innovation in its way of looking, even if that is a little destabilizing" (2015: 114). Guattari embraced dissensus and faced chaos head-on.

Guattari's voice is in some limited respects remarkably Mumfordian. At the end of *The City in History*, Mumford (1961) shows an overwhelming concern with the ecological consequences of nuclear and bacterial arsenals for the global ecosystem. His concerns cannot be diminished as only material-environmental. Of course, he does not share Guattari's love of machines, instead, for instance, despairing over interplanetary travel as escapist; obviously he would have considered McLuhan's notion of the moon as a suburb as jejune! Still, Mumford revels in the prospects of "new institutional arrangements" and "social creativity"; he rejoices in citizen-driven initiatives for "self-knowledge, self-government, and self-actualization," for the cultivation of "care, culture and love," too. Mumford would in the end recoil at Guattari's understanding of the planetary city as a vast machinic phylum, a continuum in which the military played a historic role along with the emerging city machines that stratified the phylum and have repeatedly survived the violence of the former. Enslavement is rejected by Mumford (1961: 575–6) in favor of a very un-Guattarian transcendence of "cosmic processes and eternal values," yet they express responsibility for future generations, a point that Guattari also emphasized in "negotiating the present in the name of the future" (1996a: 271). Nevertheless, perhaps Guattari had Mumford in mind when in one of his last publications he identified those who, much to his regret, still "frequently continue to oppose the machine to the human spirit" (1996: 267).

When Guattari wrote of "the city" he often had specific examples in mind, namely, his home in Paris, but also Tokyo and São Paolo. It is perhaps easier to grasp his recourse to the "network" figure if a few qualifications are added: that the network of nodes and lines is multidimensional, or perhaps multiscalar; yet visibility and functionality fall short of Guattari's insights into the complexity of urban experience. This is how architects understand Guattari's urban eco-logic, for instance, architect Helene Furján, as a subjectivation machine engaging global and molecular dimensions (2008: 297). To which may be added Guattari's sense of the spatial and temporal dynamics of urbanization, conceived as web-like, but in terms of differential speeds of technical, scientific, and aesthetic advances on a planetary phylum. Guattari (1981: 39) was once asked about his ideal city, and he replied by noting his experiences in Japan. He viewed the emergence there of "machinic mutants ... for the best and for the worst! You asked me how I see future cities, ideal cities? From the perspective of the best noted above." Obviously, this is no simple valorization of unambivalent machinism, despite Guattari's obsessive attachment to machines of all kinds, but a recognition of how urban environments subjectivate animistically, absorbing, calling, manipulating their denizens concretely and abstractly, corporeally and incorporeally, from the smallest details (i.e., a hallway) to the largest infrastructural features (i.e., bridges, highways, skyscrapers). These highly charged features are partial enunciations that catalyze the amassing of components into a consistent assemblage of subjectivation, with all of the nuances of a polyphonous and emergent formation. Within these infrastructural and, of course, architectural transferential phenomena, Guattari considers pathic apprehensions of partial enunciators, and provides as the simplest spatial example that of an ambience, which is without mediation and without reference to distinct, parceled information: for example, "as soon as one enters certain primary schools, one feels anguish oozing from the walls" (Guattari 1993: 146). Further, Guattari (2015:82) described the overall objective of Japanese architect Shin Takamatsu's practice as "arriving at a point where the building becomes a nonhuman subject capable of connecting with individual and collective human subjectivities." Such partial enunciators infuse the material infrastructure, and their pathic apprehension draws transversal lines between them, sometimes slowly, sometimes in a flash, perhaps as the institutional paint applied to cement blocks in a barracks style school is mocked by the traces left upon it by the ball games played against it and scuffs from skateboard tricks. The "objectity" of the partial enunciators entails that subjectivation, at least its extra-human parts, is objective; or, that objectity and subjectity overlap. Guattari's renewal of Mumford can be put succinctly: megamachinic animism redefines the city as a field of partial-enunciators and hence proto-subjectivations at all levels (Melitopoulos and Lazzarato 2012).

Looking back on the skirmish with meaning in information theory from a Guattarian perspective, it is evident that the exclusion of meaning is a rejection of authority and an assertion of independence. From the outset, information theory separated the transmission of information by physical signals from meaning, reducing the latter to a "subjective factor" (Cherry 1966: 44), yet meaning crept back in as information was semanticized, not only by Weaver, but by many others. Another good example is how Norbert Wiener (1967: 128) grappled with the problem of semantics in communication by building a filter into the transmission line which would be an "activating mechanism" triggering understanding or appreciation, and thus a control on meaning loss; this was a superimposition of a vague flow-through device to mechanistically filter impurities or "insignificance" out of information. Although the rejection of meaning in the transmission model has been considered in various ways "vulgar" and reductionistic within contemporary communication theory (Maras 2008), Guattari revalorized and positively expressed this deficit in a novel semiosis adapted to the world of machine-to-machine communication in a post-human landscape stripped of its ontological dangers yet not at all divorced from ecological considerations. Of course, Guattari never rejected meaningful signification outright, but decentered it in an acute criticism of the alignment of its prerogatives with dominant significations and human subjects. Guattari did not restage what is sometimes popularly described as "the sacrifice of meaning" (Gleick 2011: 416) in information theory. Rather, he made meaning a crutch that, tied to existing, compromised conceptions of subjecthood, semiological regimes, and representations, could be set aside because of the opportunities he found in machinic enslavement and the important roles played by a-signifying semiotics therein.

2

Jean Baudrillard's Anti-semiology

One of the most robust examples of a coherently elaborated anti-semiology is found in the early theoretical work of Jean Baudrillard. Baudrillard's contribution to critical semiotics is an anti-semiology that is focused on core tenets of the science of semiology, in particular the linguistic theory of value, arbitrariness of the sign, and the very unity of signifier and signified. He takes aim at the "metaphysics of the sign" in his collection of essays *For a Critique of the Political Economy of the Sign* (1981). The so-called scientific postulates (1981: 148) of what he refers to as semiolinguistics harbor metaphysical principles that Baudrillard exposes as answering to a logic that structures, once suitably leveled, both commodities and signs, and which is counterpoised to a radical anthropological process, symbolic exchange, that he valorizes in a number of ways. While the very name of symbolic exchange might entangle this anti-semiology in the web of surveys of "signs and symbols" and novel definitions of the latter marked term that have been a long-standing interest of semioticians (see Eco 1982), symbolic exchange is supposedly, however, an antidote to semiosis and code-governed structuration.

My task in this chapter will be to examine in detail the principal claims made by Baudrillard about the failures of political economy and semiolinguistics and how he uses a pseudo-algebraic process to build the terms of his critique. I will also introduce how he positions symbolic exchange in the breach. Symbolic exchange is not my main interest in this chapter. Baudrillard's recourse to symbolic exchange has been referred to by Jean-François Lyotard as a "myth," "fantasy," a "lost referent" regained positively, and an "alibi" (Lyotard 1993: 106–7). I will devote a later section of this chapter to explaining how Baudrillard derives symbolic exchange from the gift theory elaborated by Marcel Mauss for the purposes of contrast with the details of his critique of "the political economy of the sign" (eventually, by the time of *Symbolic Exchange and Death*, Baudrillard referred to this idea as "makeshift"; 1993a: 7). Baudrillard captures the homology of commodity and sign under the structural law of value, emphasizing the second principle of Saussure (a coin that may be compared with other similar coins in a given currency or with other currencies). For Baudrillard, the "sign" becomes at best "allusive"; it is the principle of value that is important, since signs are not longer exchangeable against the real as structural play survives referentiality and is imprisoned in what semioticians refer to a world of sense rather than reference. As for his reading of Mauss, Baudrillard reads Mauss against himself, developing a supercharged conception of the counter-gift, a hyperbolic deployment of potlatch, and attempts to find in contemporary society forces that compel exchanges to complete their circuit, not that of Marx's commodity-money-commodity but giving-receiving-returning.

Critique of the political economy of the sign

Not only does Baudrillard attempt to mount a critique of the political economy of the sign and the submerged power relations of semiology, but he does so at the cusp of turning away in his own work from the critical use of structuralist categories of description that played important roles in his theory of consumer society in the late 1960s. This turn prevents readers from conflating his descriptive use of typologies and categories of relationality borrowed from semiology as a statement of position. Rather, as Baudrillard's descriptions of objects thicken, the contrasting role they play with a slowly emerging alternative account is key to understanding his theoretical project. Instead of valorizing structuralist method, he builds on the back of the conceptualization of differences another account of the object, not as sign, but as gift, which underlines the failure of the object as sign: "The object become sign no longer gathers its meaning in the concrete relationship between two people. It assumes its meaning in its differential relation to other signs" (1981: 66). Meaning is decontextualized both functionally and in terms of personal bonds. The logic of signification pertains to signs whose operationality is defined by differentiation and which express status and prestige relations (situated in a coded hierarchy of objects of consumption) rather than a symbolically charged, singular object like a wedding ring that cements a publically acknowledged bond between persons. The first two contrasting logics appear thus:

Logics of signification, part 1

Logic	Operation	Object	Value
Gift	Ambivalence	Cruel sign	Symbolic exchange
Status	Difference	Linguistic sign	Sign value

In contrast to the sign, the gift is a symbolic object whose operational relations are defined by ambivalence in a psychically charged personal relationship—not between object-gifts, but between persons. Gifts are inseparable from personal relations, and it is this power that prevents their reification, their sign-ification. Gifts are "not codifiable as signs" Baudrillard (1981: 65) insisted. Sign-objects are, for Baudrillard, nothing; it is useless to describe them empirically: "An object is not an object of consumption unless it is released from its psychic determinations as a cruel or brutal sign; from its functional determinations as instrument; from its commercial determinations as product; and is thus liberated as a sign to be recaptured by the formal logic of fashion, 'by the logic of differentiation'" (1981: 67). Building upon the gift-sign contrast, Baudrillard adds the functional object defined by its use value, its logic of utility is practical, and its objecthood is that of an instrument; and economic exchange value (EcEv) of a commercial product or commodity, whose operational definition is that of equivalence within market or exchange relations. Ultimately, the latter will prove to be ubiquitous and the former vanquished. The latter two categories of logic of signification, use value and exchange value, will form part of Baudrillard's critique of the political economy of the sign, to which I will turn momentarily. It is important to note at this early stage that all of the nongift categories will collapse together as Baudrillard mounts one overriding contrast between all logics of signification and symbolic exchange, his master anti-semiological category, composed of poetic anthropological extrapolations. His choice of the widely deployed concept of "symbolic" overcomes his own misgivings: "... the category of the symbolic is a bit worn, there are too many misunderstandings over the term," as he once admitted (1993b: 57).

Nevertheless, Baudrillard persists in characterizing his symbolic relationally in terms of a series of failures and mediocrities that befall semiological thought. To the extent that semiology attempts to "save meaning and to produce meaning" it is, for him, evangelical, of which it is apparently somewhat aware (1987: 43). Semiology attempts to save meaning by means of distinctions that "domesticate" it within the terms of sign systems (1987: 44) and with regard to a basic set of binaries (signifier–signified; sign–referent). More robustly, however, Baudrillard pits an energetics of symbolic eruption against a dynamic of semiological reduction: "it is the semiological organization itself, the entrenchment in a system of signs, that has the goal of reducing the symbolic function" by watering down ambivalence for the sake of distinctive oppositions: his examples include the subordination of the consequent to the antecedent term (sunny vacation/solar excess) as the two weeks of sunshine demanded by vacationers as opposed to so-called poor weather; holiday sun is thus stripped of its symbolic force, as Georges Bataille (1991) maintained in *The Accursed Share*, by aligning solarity with an exuberance that must be acceded to in the name of loss; fundamental ambisexuality is reduced in the masculinity/femininity distinction to a specified sexual segregation, highly phallic and aggressive; and in the unconscious–consciousness dyad the former is individuated, rediscovered, and self-controllable like a piece of private property that can be surveyed and fenced (1981: 98–101).

Baudrillard deploys a brutal, sure, and cruel sign, delinked from arbitrariness and embedded in an "inescapable reciprocity" (1993a: 44) somewhat enigmatically as an escape from structure, meaning, and communication. Subjectivity is squeezed between the Manichean distinction that organizes Baudrillard's thought: symbolic and semiological. The symbolic domain includes play, seduction, and duel/challenge; the semiological is the domain of communication, structural difference, and simulation (models and codes). The symbolic rules of the game are contrasted with the laws of conduct. Wagers are opposed to contracts. Yet across this proliferating divide one finds not so much bridges as correspondences that suggest structural oppositions of terms and further dualisms of thought—for example, the fragment and the fractal; metaphor and metonomy. These correspondences suggest that just as there are orders of simulacra in the realm of signs and simulacra, there likewise may be gradations within the symbolic (i.e., different levels of metabolic communion, various energies of counter-gifts). Paul Hegarty (2004: 75) argues convincingly that symbolic violence is diminished in a culture of mediation and negotiation. This makes symbolic exchange the "omni-absent" term. Yet it does not disappear entirely. Baudrillard writes of a "minimal symbolic cycle" in which power seduces by means of reversibility, a symbolic force par excellence. The distinction between power and counterpower does not become antagonistic without a seduction that cancels them. Perhaps the paradoxical result of this pronounced absence is the foregrounding of the opposed semiological terms, so that Baudrillard's criticisms of the simulacral devaluation of semiology in the disenchanted signifieds serve as prophylaxes against the real and erase any sense of obligation to it. Baudrillard (1981: 82) puts a great deal of emphasis on the effects of signifier differentiation and this allows him to link meaning production with the systemic inducement of needs, neither of which have their origin in an individual subject. What he calls a semiological reduction is for him ideology, and it is worked through distinctive oppositions that feign neutrality but actually discriminate between their terms: "Signification does not always carry discrimination with it (phonemic differences at the level of language) but discrimination always presupposes signification—the sign-function that reduces ambivalence and the symbolic" (1981: 101). Further, it is the abstract character of semiological systems and their dichotomies (cleanly coherent) and their autonomous status that gives to them a total, and thus, ideological, power.

Baudrillard thinks of the sign as a hieroglyph, both in cultural context and in theory. The logics of status and market are bound together by Baudrillard because economic exchange value manifested in a purchase which is often sustained by a re-appropriated use value invests the commodity with a sign value; there is also in this process a kind of "transeconomic" element of expenditure, destruction, and "sacrifice" beyond parsimonious investment. This latter element is glimpsed for Baudrillard in the art auction (1981: 113) and it is this cultural context in which he invests a counterflow: not the transmutation of all measures of value into economic exchange value, but the opposite: the transmutation of economic exchange value into sign value: "This is the process of consumption considered as a system of sign exchange value [SgEv]" (1981: 113).

Logics of signification, part 2

Logic	Operation	Object	Value
Utility	Practical applications	Instrument	Functional use value
Market	Equivalence	Commodity	Economic exchange value

The logics that express this transmutation from EcEv to SgEv are designed to "explode" the use value/exchange value distinction and generalize political economy into that of signs, into the production of signs on par with any material production and therefore not merely superstructural. This "integration" of signs into political economy is Baudrillard's brand of semiocapitalism, and it requires a specific sumptuary operation surpassing both the social labor that went into sign production and the economic value that defines commodity exchange: the mastery of accumulation gives way to the mastery of sign values and of signification through expenditure as a kind of "semiotic privilege" (1981: 116). Privileging mastery of signification (i.e., within a caste of collectors the ownership history of a work bestows legitimacy) over the ownership of the means of production injects a semiotic dimension into class logic and redefines material production itself. Baudrillard's analysis hinges on his transformation of money from economic to sumptuary value and aristocratic parity among collectors. There is no place for symbolic exchange in this process, except to note that it becomes a satellite of sign value (even if buying a painting can be rationalized as a good investment or a particular work, like a Van Gogh or Gauguin, can be thought of as embodying universal aesthetic values). These are simply alibis

(1981: 120). The only truly symbolic aspect of purchasing a valuable work of art would be in destroying it, that is, to become intoxicated with the prestige that such a loss would bring. This mysticism of expenditure is borrowed by Baudrillard from Bataille as a foil.

If the art auction served as a "nucleum of the strategy of values" (1981: 122), for Baudrillard the proof of concept is in a general conversion table of values that transcends particular sites and hypothetically marks transits between four fields.

Phase 1: Hypothetical general conversion table

Use Value, UV	1. UV	EcEv
	2. UV	SgEv
	3. UV	SbE
Economic Exchange Value, EcEv	4. EcEv	UV
	5. EcEv	SgEv
	6. EcEv	SbE
Sign Exchange Value, SgEv	7. SgEv	UV
	8. SgEv	EcEv
	9. SgEv	SbE
Symbolic Exchange Value, SbE	10. SbE	UV
	11. SbE	EcEv
	12. SbE	SgEv

Each field has three transits. Across the fields there are unities (1, 4) derived from political economy such as the "direct unity (Marx 1972a: 41) of exchange value and use value in the commodity, as well as transfigurations (2, 5) of forms (commodity-object-sign), in addition to which there are transgressions (3, 6; 9, 3) of economic and semiological relations.

The first field concerns UV:

1. UV-EcEv: the value in use of an object, taken in terms of its distinct properties or contents, and the individual needs that it satisfies, enters the field of economic relations when it expresses itself as exchange value. Becoming a commodity: what was a particular use value of something for oneself enters into exchange relations when something is produced for somebody else and exchanged.

2. UV-SgEv: the utility of a thing is destroyed by forms of consumption that display their conspicuousness and/or unproductiveness (i.e., bohemian idleness). Here, the alienation of a commodity as a use value—henceforth inexchangeable and disconnected from other use values—is assisted, according to Baudrillard, by advertising that stages a "'cultural' system of differentiation" (1981: 124) with dubious virtues (i.e., semiautomatic versus fully automatic appliances) by confounding functional with statutory differences. Embedded in this transit is an entire critique of needs as systemic effects that cannot be pinned down to specific objects that satisfy them.

3. UV-SbE: this is not a semiotic reconstruction of use value but an *an*economic destruction of use value in the form of a truly sumptuary destruction (violent and frenzied) of riches or potlatch; nonproductive expenditure or the squandering of riches is a social act witnessed by the "receiving" group that allows those capable of loss to acquire prestige (Bataille 1991: 69–70). The salient distinction here is between *consommation* (consumption) and *consumation* ("pure and simple destruction of goods"; Baudrillard 1998: 43). Symbolic value is always "something more" than a thing possessed, a useful, and rational object, and expenditure trumps accumulation and similar cold calculations.

In the field of economic exchange value, the fourth transformation is the first in reverse.

4. EcEv-UV: a commodity is and is not a use value, Marx states, in that it is a use value if it satisfies its owner's needs immediately, yet not a use value to the extent that it is a means of exchange and may then be established as a use value for others. This movement of alienation is still outside of sign exchange value yet inside political economy in general. Baudrillard (1981: 124) qualifies this "transfiguration" in two ways: as a consecration of exchange by use value, suggesting Marx's remarks on the ways in which commodities realize use values

as exchange values; and as the route for a metamorphic change of form from commodity to object, defined semiologically.

5. EcEv-SgEv: read together with 2 above, it marks out the consumption of differences in the field of sumptuary value. The economic is transfigured into the semiological; commodity becomes sign form. As Baudrillard explains, "to become an object of consumption, an object must first become a sign ... it is thus arbitrary ... it derives its meaning from an abstract and systematic relationship to all other sign-objects" (1996: 200). And further, "if it has any meaning at all, consumption means an activity consisting of the systematic manipulation of signs" (1996: 200). Ultimately, in the field of consumption, there are no more objects, only signs that cohere in a system of organized manipulable elements constituting an ambiance (a designer ensemble of balanced tonalities and signified temperatures and moods).

6. EcEv-SbE: taken with point 3 above, this transit marks a transgression of commodities and sign forms into symbolic exchange that has no code of "value": "All forms of value (object, commodity or sign) must be negated in order to inaugurate symbolic exchange. This is the radical rupture of the field of value" (1981: 125). It necessitates a second phase of the process outside the above transits in which the collapsed theories of value oppose the field of symbolic exchange. Baudrillard does not explicitly query the labor theory of value in the transit from EcEv into SbE yet presumably the human labor embodied in a commodity would not carry forward into aneconomic exchanges.

7. SgEv-UV: still, reconversion is entailed from the sign to the materially useful thing that does not embody labor time. Back to found value, the satisfaction of one's own needs and the so-called incomparability of use values. However, Baudrillard believes that use value is an untenable category and he finds in it a number of abstractions, as Mike Gane astutely pointed out (1991: 87): (i) a common denominator of rational-functionality, hence a code of utility; (ii) illusory role of the private individual, a hangover from bourgeois economy; (iii) the critique of needs as a human endowment (naturalist ideology).

8. SgEv-EcEv: a reconversion (of point 5 above) and a further example of a ceaseless cycle of exploitation from cultural (code) to economic (capital) and back again.

9. SgEv-SbE: taken together with points 3 and 6 above, the sign is deconstructed and transgressed.

10, 11, 12: SbE-UV, EcEv, SgEv: the reconversion of points 3, 6, and 9 above. Baudrillard characterizes this reconversion as a single process, yet it displays various facets of reductionism, metaphysics, and abstraction.

1–12: the matrix of values is characterized by Baudrillard as preliminary, its results are uneven, the operations of the logics of signification are difficult to "articulate clearly," and the combinatorial possibilities are "merely formal symmetries" (1981: 126). Hence, a second phase is required in which the equations are a little more compact, and less mechanical and matrixial.

In summary, the transits that Baudrillard emphasized are:

Logics I	Logics II	Logics III	Transitions	Descriptions
1	4		Commodity-object	Classical and Marxist political economy
2	5		Transfiguration	Economy-social caste privilege
3	6	9	Transgression	Deconstruction of the semioeconomy
5	8		Reconversion	Cultural privilege-economic privilege
10	11	12	Inversion of 3/6/9	Breaking and reduction of symbolic exchange

Phase 2: Homologous relations and discontinuous equations

Baudrillard then asks himself how he might "extract some dominant articulation" from the table. He produces the following equation:

$$\frac{\text{SgEv}}{\text{SbE}} = \frac{\text{EcEv}}{\text{UV}}$$

Read vertically [upward], the relation between symbolic exchange and sign exchange value is based on reduction, abstraction, and rationalization, which is also evident in the deconcretization of use value into exchange value (embodiment of human labor time in commodities as an abstraction). In short, the processes in play on both sides of the equation are formally identical. The larger economies at issue are semiological (sign value) and Marx's critique of political economy, yet Baudrillard's project is to launch a critique of the political economy of the sign, that is, an anti-semiology whose positive content is a theory of symbolic exchange; thus, his version of semiology situates it like bourgeois political economy before Marx's critique. And, ultimately, Baudrillard situates himself ironically as a new, aberrant sort of "Marx" with his sights set on semiology, not on political economy proper. Importantly, the form and not the content makes this homological argument possible: "If the political economy of the sign (semiology) is susceptible to a critique in the same way as classical political economy, it is because their form is the same, not their content: sign form and commodity form" (1981: 126). For Baudrillard, the focus on homological relations of forms is a "considerable advance" over the matrix, yet the vertical reading must be reinforced by a horizontal reading. Hence, in a preliminary way, a second equation:

$$\frac{\text{SgEv}}{\text{EcEv}} = \frac{\text{SbE}}{\text{UV}}$$

Wherein sign value is to economic exchange what symbolic exchange is to use value. However, this is not "true." While the two forms (sign and commodity) are mutually implicated, this is not so for symbolic exchange and use value: SbE transgresses UV and UV reduces SbE. "The formula is then not coherent," Baudrillard surmises (1981: 127), because symbolic exchange has no place among the other values and, indeed, opposes them. So, the equation is not an identity and not provable. As Gane puts it, "the formula ends in incoherence" (1991: 84). Is this a staged failure? Perhaps. After all, recourse to homologous structuration as a principle is well embedded in semiotics as a way to demonstrate equivalences among systems (i.e., as attempted by Rossi-Landi in his modeling of the homologies between linguistic and material production, 1977). And the obvious objection to Baudrillard's gambit—that he is padding his anti-semiology with semiostructural maneuvers—goes quiescent rather quickly.

The radical heterogeneity of symbolic exchange "bursts the formula" and precipitates a new equation:

$$\frac{\text{EcEv}}{\text{Uv}} = \frac{\text{Sr}}{\text{Sd}}$$

EcEv is to UV what the signifier is to the signified. The vertical coherence is in this equation joined by the horizontal affinity of EcEv and signifier and UV and signified. This is Baudrillard's semiological reinscription of general political economy in a homologically structured global relationship that telescopes the final equation (see below) inasmuch as symbolic exchange has been expelled from the field of value. The process initiated by Baudrillard at this juncture may be best described as a pair of bar games.

First, there are bars of logical "implication" between EcEv and UV and Sr and Sd; additionally, the relations between EcEv and Sr and UV and Sd are called "structural," by which Baudrillard means inclusive and constitutive of a unity, like Saussure's (2011: 113) image of language as the recto and verso of a sheet of paper. These are weak bars that operate throughout the field of value.

Second, there is a strong bar that separates and is a "line of radical exclusion" (1981: 128):

$$\frac{\text{EcEv}}{\text{Uv}} / \text{SbE}$$

All of the equations boil down into this one great exclusion between fields of value and symbolic exchange: a critique of the former, focusing on the linguistic theory of value, use value, and signifier fetishism, requires a theory of the latter which would be a "revolutionary anthropology," derived from Bataille and Mauss. What exactly is proposed by this alternative account? Baudrillard borrows heavily from a number of anthropological studies to flesh out symbolic exchange, but these matters will not concern me here (see Genosko 1998). My focus is on the critique of semiology in its own right. To this end, I want to turn to Baudrillard's analysis of the impurity of structural relations in his discussion of use value.

"Use value is the expression of a whole metaphysic: that of utility" (1981: 133), writes Baudrillard, separating:

$$\frac{\text{abstract and general}}{\text{concrete and particular}}$$

but, at the same time, arguing that despite Marx's efforts to maintain a strong bar between the upper logic of equivalence and the lower incomparable and "determined valence" of utility failed since use value is already caught up in a rationalist and abstract, yet naturalized, relation between an object and the individual's needs it satisfies. Utility is in short a "rational-functional common denominator" (1981: 132). Baudrillard reserves a special place for goods that are exchanged symbolically; they alone are incomparable. The idea of an object with an innate function is, for him, "false evidence": it is "an anthropological illusion that claims to exhaust the idea of utility in the simple relation of a human need to a useful property of the object" (1981: 132). Use value is a social determination and there is also fetishized social relations of use values. Therefore, the logic governing EcEv and UV is the same. Baudrillard then returns to a familiar equation:

$$\frac{\text{EcEv}}{\text{Uv}} = \frac{\text{Sr}}{\text{Sd}}$$

The structuration displayed here is not very symmetrical and neutral. Why, because structural logic contains advantage and disadvantage, in this instance, exchange value and the signifier have strategic positions over the lesser tactical value of use value and the signified; the latter do not have the "same weight" as the former. Hierarchy slips in and the lower forms become only "effects" of the upper while providing "guarantees of the real" (concrete, objective; 1981: 137) for the uppers. Use value and the signified provide "alibis" of the real; indeed, use value provides the alibi of equality (democracy of needs) before objects taken as use values and the eternal rather than historical inscription of "the useful appropriation of objects by man in need" (1981: 139). Again, the lower terms are "satellites" of the uppers and thus lack autonomy and can offer no alternatives. Despite Marx, Baudrillard finds mystery in use values. To which may be added, that despite Saussure's images of the unity of signifier and signified and consequences of their separation (either pure psychology or phonetics), Baudrillard finds in the absence of "ontological autonomy" (independent determination) between acoustic image and concept a conflation of signifier and sign (a notion Saussure once harbored; see Bouissac 2010: 34 and 93) to the marginalization of the signified. Hence, the lower completes the upper:

$$\frac{\text{dominant form}}{\text{submissive alibi}}$$

Perhaps, from the critical perspective I have been displaying, Roy Harris (1988: 126) moved in the right direction when he claimed that Saussure was worried that his readers would consider his conception of the sign a "theoretical abstraction." Even its most concrete (empirical) part is highly abstract according to Baudrillard's arguments. Perhaps in this matter Saussure's concerns were well founded.

This focus on ideology and the signified is still too traditional for Baudrillard. Ideological masking of a connection with the real through submissive terms does not capture the thorough, internal form of ideology across the logics of the commodity and sign forms. A slash (virgule or solidus) of implication stands between EcEv and (not directly over) Sr and UV and Sd. For Baudrillard,

> This is the functional, strategic split through which the form reproduces itself. It signifies that ideology lies already whole in the relation of EcEv to UV, that is, in the logic of the commodity, as is so in the relation of Sr to Sd, i.e., in the internal logic of the sign (1981: 144).

Ideology is everywhere! Both sides of the slash are perfused with a logic of abstraction and exploitation because the commodity inhabits the sign and the sign inhabits the commodity. Both material and immaterial contents alike are exploited by "the code" that determines "the interplay of signifiers and exchange value" (1981: 146). Equivalence thus reduces ambivalence, a symbolic force, in so-called rationally regulated circulations and combinations of commodities and signs, under a general form of political economy (called the "object form"), and neither separately nor specifically.

The critique of the sign's second term (Sd) as an alibi and satellite of its first term (Sr) goes further than disrupting the claim that the relation between the signifier and signified is arbitrary. It cuts to the heart of semiolinguistics. Suffice to say that Baudrillard does not concern himself in depth with Saussure's (2011: 133) observation of absolute and relative degrees of arbitrariness along a continuum between two improbable extremes. Nevertheless, Baudrillard does override the pertinence of the insight that symbols, in Saussure's understanding of them as signs that are "never wholly arbitrary" (2011: 68), are subject to the code of equivalence and therefore it is not relevant for him that a relationship between signifier and signified may be partially analogical or natural. The foundation of Baudrillard's critique of semiology is the metaphysics of equivalence: it overrides any questions of relative and whole arbitrariness by pointing to the "very fact of positing an equivalence between such and such a Sr and such and such a Sd" (1981: 149). This positing of the unity of sound image and concept, a correlation of two sides each "equally discrete," is a brand of rationality that makes the sign a "discriminant": it banishes or severely limits by reducing, as noted earlier, anything inexact that might upset its exclusivity. Looking at signs in this way as "one-to-one" correlations leaves room for equivocal, multivocal, and even polyvocal relations, as long as the principle of self-containment is neither upset nor challenged. Meaning is never beyond the control of signification (and value) and the constrained combination of terms. Ambiguity is tolerated because it is harmless. Beyond equivalence there is ambivalence and the "rupture of value ... and the emergence of the symbolic" (1981: 150). Baudrillard places a great deal of emphasis on the sign considered as a totality, a positive whole, expressing signification. Here, he has less to say about value and interdependence, but opposition and negative differentiation require linguistic terms: "While retaining their discreteness, Sr and Sd are capable of multiple connections" (1981: 150). Both pillars of the linguistic theory of value (exchange of similar and dissimilar units) are implicated in Baudrillard's critique of semiologic. Any rupture of such semiologic would entail the critical "resolution of the sign" [and its equation] and, in short, the "abolition of the Sr and Sd" (1981: 150).

There is nothing in semiologic that can resolve the equations in the field of the political economy of the sign; it has to come from outside, beyond the bar, and forcefully break in.

Further to the challenge issued to the sign's arbitrariness, Baudrillard expands the field of critique by including Emile Benveniste's displacement of arbitrariness onto the sign–referent relation (reference field) and his rephrasing of the arbitrary relation between the signifier and signified as governed by necessity (sense field). Arbitrariness is not, for Baudrillard, the opposite of motivation, relative or otherwise; it concerns the "exact correlation" of signifier–signified: "arbitrariness lies in the 'discretion' which alone grounds the possibility of the equational relation of the sign" (1981: 149). Displacing this "metaphysics" does not solve anything as, thinks Baudrillard, the scission is actually between signifier-form and signified-content including reference (in short, thought-real). This move entails that the referent is not external to the sign: "its only reality is of that which is ornamentally inscribed on the sign itself." In a profound sense, the referent is the "reflection of the sign" (1981: 151). The adequation of the signified and the real is, albeit naïvely, accepted by speakers, claims Benveniste. Any effort to as Baudrillard puts it rationally "carve out" abstract separations and designated types of "false" distinctions between discrete parts is an exercise in futility that masks the fictional difference between signs and the real. Yet Baudrillard has himself to borrow a new entity, the signified-referent (rft), the "shadow" that haunts the signifier: "the Sd-Rft is a single and compact thing, an identity of content that acts as the moving shadow of the Sr. It is the reality effect in which the play of signifiers comes to fruition and deludes the world" (1981: 152). Thus, the psychical nature of linguistic signs is key to Baudrillard's critique, as any effort to build bridges with referents, like that of Benveniste, only results in the formulation of a "referential rationale," an alibi of the real (as already designated). His signified-referent illustrates the "frictionless space" of the conceptual and perceptual (1981: 155). What he is intent on displaying is the metaphysical concept of motivation/arbitrariness and how it attempts to miraculously join what the semiologic has modeled asunder, that is, posited as separate: "Concepts are quite meaningless when they are busy bridging non-existent gaps" (1981: 154). The referent, as ornament of the signified, is the shadow of the sign that cannot be jumped over. The collapse of the referent into the signified is Baudrillard's illustration of this impossible situation: you can't jump over your own shadow and somehow reach reality, since the horizon of the sign is the shadow it casts as a referent and beyond which it is impossible to step. As the anti-semiological route into Baudrillard's theory of simulation, this critique has an important role to play in his overall intellectual project. I will now provide an example of how this theory works with respect to a cultural signifier (margarine) and its supposedly real referent (butter).

Realer than real: The simulation of butter

Barthes decoded the mythology of margarine as a rhetorical species of inoculation. He described the promotional demonstration of margarine's blemishes as a carefully cultivated virtue in a cultural vaccine that selected contingent over essential evil. Inoculation is one of seven principal rhetorical figures that Barthes develops to analyze the duplicity of bourgeois myths. Margarine's commercial valorization takes place on the grounds of its imperfections, as well as its secondariness in relation to butter. That is, margarine's resemblance to butter entails that its simulacral features themselves become the foundation of its advantages, and of its delights in the mouths of skeptics, despite their traditional dairy allegiances. The homeopathic vaccine of inoculation against greater evils clears the way for its widespread acceptance.

Readers of Barthes's original essay (1957: 44) on this point, "L'opération Astra," may wonder about the terms of reference of his analysis since the English translation exposes far too little of the focus on a specific French brand: "Astra." Barthes's margarine was the leading French brand (a product of Dutch agribusiness giant Unilever's French subsidiary Astra, after which the product was named) and not a generic substance in a plastic tub. Indeed, the mythological strategy of inoculation was also found to apply to plastic. Plastic has a lot in common with margarine as a mythologized substance—triumphantly chemical, smooth, and shiny. In Barthes, mythologies are nestled within mythologies—the essay's title is translated nonspecifically ("Operation Margarine" and not "Astra";

1973: 45) within the general rhetoric of an emerging promotional demonstration applicable to a variety of substances. The referent recedes from view in translation.

However, it was Barthes who put margarine on the table. Barthes taught that margarine is before all else a relational substance that speaks another name: "butter." This is precisely what you can hear, if you listen closely, when the lid on a tub of margarine is lifted: "butter." And of course for Canadians and Americans of a certain vintage this is also the promotional discourse of Parkay margarine of ConAgra whose "Talking Tub" muttered "butter" as an act of provocation, and has been nattering since 1973. Of course, today margarine speaks butter's name with a "not"—"I Can't Believe It's Not Butter"! What is the calculus of simulation that animates margarine's history and carries its mythologies into the present? Margarine is simulacral and its history and destiny plays out this role in degrees of "likeness" to the natural, original reference point of butter. Every attempt to justify the distinction between margarine and butter along a continuum of simulation and reality invokes metaphysical concepts. Following Baudrillard, you cannot transcend the muttering of margarine and regain the real thing, butter. The "naturalist guarantee" that butter enjoys is provided by an unreliable source.

Butter—simplifying a materially complex polycultural substance to be sure—is destabilized as margarine's status as a counterfeit or fake is superceded. Margarine turns the tables on butter not as it achieves perfection in the erasure of difference with butter, nor through butter's exhaustion, for example, of legal means that prevent margarine's drive for similarity, but as simulation perfuses the relational calculus of these food objects. The game of appearances, if you will, of the unstable separability of margarine and butter, which both have played out since margarine's creation as *buerre économique* in 1870 in France by Hippolyte Mège Mouriès, becomes undecidable. This is a nonspecific process of contamination. I do not want to attribute this mutual metabolization to butter's waning mystique nor to margarine's triumphal technological prowess of reproduction. Neither am I adopting an explanation based on fatigue produced by legislative differentiation in conjunction with the lifting of codes protecting butter's integrity, and hence opening onto a profligate indifferentiation; and I am not playing the health card of shifting analyses of nutritional value (margarine's surging

Mediterranean profile and butter's sinking fattiness) that contributes to the pair's implosion. To put this slightly differently: what if margarine is no longer obviously secondary to butter? And, what if butter is no longer primary in relation to its pretenders? Simulation's effect is that such ordering cannot be upheld.

Both margarine and butter issue from models. Only a critical dredging operation in the histories of these substances can reveal the extent of butter's strategies for "real-ization" (Hegarty 2004: 51) that situates it in simulation. Still, the great chain of modeling is so long and intrusive as to go unnoticed as models are produced from market research, changing production processes, trends in artificial and natural additives, right into the mouth of the consumer with the idea of "mouthfeel aroma," a point made about bread, but one equally relevant here, by Victoria Grace (2000: 107). Butter is realer than real, yet it remains resistant to this claim because it carries with it nostalgic obligatory symbolic attachments to traditional activities like dairying, to the necessity of milk. All of butter's attributes are gathered under this rapidly eroding sign: nature, farming, processes like milking, creaming, and churning; richness, purity, and uniqueness. Let's listen to Margaret Visser (1986: 101–2) describe butter's superiority. Writing of French *crémeries*, she observes,

> Butter appears packaged in the now-usual silver or gold foil, but often the very best of it is still served from the motte—a huge shameless tower of unwrapped voluptuousness, with gleaming facets where chunks of butter have been cut with a wire to the specifications of the customer. ... Butter in a motte is not squared-off, brand-named, labeled, and "industrialized": it constitutes a monumental snub to the concept of margarine.

As beautiful as it is surely false, Visser's effort at contrast does not acknowledge that butter has been packed in various ways for hundreds of years—cut in pats at home; buried in the bogs in basket-kegs; and exported in firkins, not to mention colored golden (Visser 1986: 89). By the time the proto-global butter trade emerged during the mid-eighteenth century, Irish butter was standardized in regulated barrels, its salt content legislated, prices set, and its grading (from mere grease unfit for human consumption to the finest product) soon thereafter enshrined as a brand burned into all the made-to-specifications wooden firkins in the famous Cork Butter Exchange in Ireland (Rynne 1998: 62). Visser's point is, however, well taken as the manner of butter's presentation is a social sign of status and its inherently desirable qualities (the naked motte, mound of unwrapped desire) are progressively packaged and cut down to size. Recall the words of Thomas Mann (1996: 30) in describing young Hans Castorp in *The Magic Mountain*: "Only reluctantly would he have eaten butter served in pats rather than in fluted little balls." Bourgeois gastronomic signification hasn't changed much as fluted little balls of butter, delicately curled and chiseled, resting in shaved ice, are a signifier of gentility in what Barthes called ornamental cookery. This distinction marks the common country pat of the farmhouse from urban bourgeois gastronomic aesthetics. Those fluted butter curls circulate widely in dairy advertising as table decorations and signify against standardized restaurant pats and other less decorative presentations.

The problem of counterfeit butter begins before the invention of margarine in the time of butter's standardization at the Cork market. Rogue butter is no match for what would be margarine's decisive intervention. The intervention is slow. The original *buerre économique* was, as historians of the substance admit, not very much like butter at all except in name. The name was forcibly changed to margarine or oleo-margarine because of the legal protection enjoyed by butter and the political influence of the dairy industry. Oleo-margarine was made from animal tallow—rendered animal fat from cows or sheep. As one historian mused: "Little imagination is needed to conjure up an impression of how Mège Mouriès' first margarine must have tasted ... no efforts were spared to improve the taste" (Boldingh 1969: 183).

Almost all definitions of margarine claim that it resembles butter and may be substituted for it. It is an alternative to butter as far as delivering fat is concerned. But at first aroma was a challenge unmet; instead, the production process of creaming directly imitated butter technology (Feron 1969: 113). Aroma is tackled gradually, as is appearance and nutritive equivalence. Margarine plays with the obligatory bond of butter and milk fat by substituting suitably neutralized vegetable or animal fats and oils. This manipulation of a key bond involves the use of soured milk in order to give to margarine a familiar butter-like and less fatty taste. It is not completely successful because not all the requisite components of butter's aroma can be achieved this way (Boldingh 1969: 109). Still, margarine strives for equivalence in a variety of ways—through the use of natural or artificial colors and flavors, where they are not subject to specific taxation or disallowed altogether; and by means of advertising campaigns that promote various sorts of comparisons with butter, within the bounds of typical restrictions on permissible terms. For much of its history, margarine has remained in the orbit of butter, recalling Barthes's cosmic brand. It is only when the gravitational pull of approximation gives way and the original reference is obliterated that margarine can pursue the pure potentiality of its generation by formulae and models. It is in an exalted simulation that margarine begins to turn the tables on butter. It is in simulation that margarine comes to truly know on which side its bread is buttered, as it were.

Margarine has struggled against its technological-industrial history in which it is linked with meat-rendering plants; it simply lacks "class," as Visser notes, due to its low cost, mass production, mass appeal, and service rendered to fat-deprived populations (Visser 1986: 112). Perhaps this prejudice is a veiled form of anti-Communism since the USSR was a great friend of the margarine industry (Stuyvenberg 1969: 308–10). Even here care must be taken to note how the lowly East German product, "Romi" margarine, was enshrined in the personal aesthetic of Joseph Beuys. For Beuys, "Romi" was one among many products used in his *Economic Values* series of related works from the late 1970s (*Wirtschaftswert Apollo*, 1977, consisted of a signed margarine container). The package was valorized by Beuys in a pop art gesture and integrated into his personal mythology as a fat source, thus taking pride of place in his personal pantheon of substances that provide energy and warmth and creativity. On the whole, however, margarine has genuinely suffered from a remarkable variety of legal prejudices and "nuisance legislation" that frame it as fake, fraudulent, and even criminal.

Cleverly "vitaminized" for the life cycle, aromatically enhanced, flexibly flavored for market niches (cheesy, meaty, or nutty), with ethnic nods (ghee, shea), margarine comes into its own. Released from the indignities of legislation that once controlled its color and threatened stigmatization—make it the color of the wainscoting in the Reichstag!—margarine achieves a measure of freedom for self-invention. But this freedom is simultaneously revealed to have been butter's secret.

To admit that margarine is "not inferior" to butter fails to break from the original reference's pull. It is merely a negative assessment of potentiality. That it is "much yellower than coloured butter" reveals a secret. Writing in the 1950s, chemical engineer A.J.C. Andersen (1954: 106) observed,

> Most countries now permit the colouring of margarine to the traditional butter yellow colour by means of added colourings and dyes. Butter is often similarly coloured artificially to ensure uniformity of colour, which if not thus standardized varies with the season and the feeding of the cows.

Butter, too, is standardized and artificially enhanced, a product of agribusiness. Its appearance is modulated in accordance with the principles of uniformity that mask its alleged naturalness and vicissitudes beyond high-tech production processes. Knowing this might make one nostalgic for the old sleight-of-hand. Butter's own strategy of the real, that is, its mediated nature, might seem to be no more than a mere dripping; yet it, too, is a real "always already reproduced." Beyond a reference to a natural substance, beyond this substance's dimension of chemical enhancement, is margarine's reference to formulae that give to it an unrivalled adaptability. As the former scientific director of the Margarine Department of the Société Astra put it: "In appearance, colour, lustre, and plasticity, margarine is superior to butter." This statement of faith in simulation is perhaps not surprising for a margarine scientist. But he continues: "The plasticity of the product can be adjusted at will ..."; it can be "tailor made." And there's more: "the nutritional qualities of margarine, too, can be modified with comparative ease... when the formula is worked out, the newest development in the field of nutrition can be taken into account" (Feron 1969: 119).

The Dutch organic chemist and fat researcher Boldingh (1969: 203) adds: "By making a correct choice of fat composition the modern producer can now adjust his products to meet current consumer habits by giving a wide spreadability range." This is industry chatter at the altar of modulated difference, a kind of giddy optimism of a product generated through formulae. When the tables are turned there is no choice between margarine and butter for we have passed beyond the counterfeit and industrial seriality: "only affiliation to the model has any meaning," as Baudrillard explained, as modulation catches up butter in a shared process.

Butter's secret is that it, too, issues from models and codes. The real referent is simulated. Butter's natural reference turns out to be artifactual and manipulable. Butter's many so-called faults are contingent upon the milk source (which breed of cow), what they ate (feed and additives), when it was produced (season), degree of exposure of the milk to daylight, etc. The color of butter is adjustable and creameries use a device called a Hansen's Butter Colour Scale to fix the "exact shade" adequate for the market at issue (a similar scale is used to modulate the color of the flesh of salmon).

Butter is colored to look more like itself, or rather, more like a coded set of attributes, and by entering into the play of marginal differentiation and combinatory possibilities becomes more like margarine. Baudrillard cites the stations along the spread out signification process from Sr to Sd/Rft, either two or three positions, as the field of play of distinctions and equivalence (1981: 157). In my example of margarine and butter, the field is traversed forwards and backwards in equal measure, by mutually realer than real substances.

In the era of butter blends spreadable straight from the refrigerator (Gay Lea's Spreadable Butter Blend with Canola Oil is described as a kind of a dream solution to an age-old problem), perhaps Baudrillard's (2003a: 15) advice should be heeded: "The charm ... of the simulacrum is that it allows us not to choose between illusion and reality." Julia Child may still signify butter for a generation overexposed to Martha Stewart and Jamie Oliver, and local farm-based butter production may be a "thing of the past," even in Ireland, but today's butter has taken on margarine's vaunted adaptability in a deregulated environment which permits hybrid modeling. The consequence is that choice is relaxed as not only margarine gets to be "better" than butter, but butter, too, is "realer" than ever before, that is, more openly simulacral and margarine-like. A day may come when one overhears: "I Can't Believe It's Not Margarine."

Staggered signification

The privilege enjoyed by the signifier and exchange value is hidden in the field of signification. This privilege is reasserted in the distinction between denotation and connotation. Across the field between the signifier and object is an adequation founded on the myth of denotation's objectivity. As Barthes explains, the expression-relation-content (ERC) system may become one element of a second system in which the ensemble ERC serves as the expression of a secondary content (Barthes 1967: 27). The primary system is denotational and the secondary system is connotational. However, when the ensemble of ERC is the content of a second system conjoined to a secondary expression, there is first an object language and second, a metalanguage. Thus, metalanguage and connotation are "mirror opposites," says Barthes. This can be put a bit differently in terms of Sr and Sd, but the effect is the same (Barthes 1967: 90). Semiotics of semiotics: connotative and metalinguistic. The denotative level "always remains," Barthes observes, and is the condition of the possibility of discourse, the carrier of connotation. Barthes refers to connotation as a "cap" on denotation, which is not exhausted in the process, while Baudrillard refers to connotation as an effect of denotation, the relation of staggering setting-up an infrastructure–superstructure-type distinction. But even this observation does not do justice to Barthes's penchant for building towers of signifying systems. For he adds another level to his tower: on the first level there is a real code (an object or image the perception of which entails action) above which is a second level of denotation (linguistic) or terminological specification consisting of sentences and propositions; and, thirdly, a rhetorical system of phrases that is "pure connotation" (Barthes 1967: 29–30). Of course, there is more to come, but reaching new heights does not change anything.

Baudrillard attacks the denotation–connotation distinction on the grounds of denotation's false objectivity and proximity to "reality." Indeed, there is in Barthes's semiology a claim for denotation's "naturalizing" of connotation (1967: 91). Baudrillard states: "denotation is never really anything more than the most attractive and subtle of connotations" (1981: 158). And he notes denotation's "illusory" first rank among meanings—telling us "something true" (Barthes quoted by Baudrillard 1981: 158). This appearance of truth-telling at the denotative level is a "false ingenuity." Therefore, the unreal distinction lines up this way:

Denotation	Connotation
Primary	Secondary
Infrastructure	Superstructure
Use value	Exchange value
White light	Visible spectrum

The allegedly primary, objective, useful status of denotation is but an effect of the play of the spectrum of connotation concentrated into a single "natural" phenomenon—an effect of the code, as Baudrillard would have it. The positions of Baudrillard and Barthes appear to align to the extent that both recognize in different conceptual nomenclatures the highly ideological character of denotation—made more so by the fact that it seems to resist ideology in its functionality and "innocence."

Signs must burn!

As we have just seen, it is the "objective innocence" of semiology that Baudrillard objects to and wishes to overcome definitively by recourse to symbolic exchange. But the route is fraught with dangers and requires the development of critical perspectives like the proverbial passage between Scylla and Charybdis, or the Sr and Sd/Rft. Neither Sr nor Sd can survive the passage or surpassing of sign exchange value into symbolic exchange as both carry with them dangers of reproducing the kinds of metaphysics Baudrillard has identified up this point. For this reason signs must burn and nothing must survive the inferno of criticism directed first at the Sd/Rft (the "idealism of the referent") and then at the Sr (the "terror" of which is defeated in the name of the "real").

Baudrillard (1981: 160) has recourse to the philosophy of Jacques Derrida, but offers few details. It is already evident that Baudrillard's identification of metaphysical attributes of the sign owes a good deal to deconstruction's exposure of how metaphysical systems force components into distinct places and define their relations (such as through arbitrariness, hierarchized pairs [speech/writing], and any center such as the subject's presence to itself). The appeal to symbolic exchange is a broadly deconstructive effort to reverse such arrangements. It "haunts" the sign, says Baudrillard, and in the respect that it subverts signs, it cannot "be named, except by infraction (*effraction*)" (1981: 161). Nothing much, maybe nothing, can be said about what is outside of signs. Symbolic exchange cannot be erected between separated terms with positive relations ("structural copulation" by the bar of inclusion, and variations on this in structuralism) that only underline their separateness. Separateness and unity are equally artificial. Efforts to regain the real or referent by means of signs erect a simulacral symbolic: "Only total revolution, theoretical and practical, can restore the symbolic in the demise of the sign and of value" (1981: 163). This uncompromising stance, at least theoretically, which is the only revolution that Baudrillard pursues, is partially in the spirit of Derrida's *différance* and the calculated discomfort of thinking it: the "a" cannot be heard; "it cannot be exposed" (Derrida 1973: 134). Neither word nor concept, it is not, like Baudrillard's ambivalence, against meaning; yet it is not for it, either, since Derrida states that *différance* cannot be understood as a sign (as a "deferred presence" that is represented). It is, rather, the condition of a number of possibilities: conceptual systems of differentiation like Saussure's; the play that produces differences; the movement (of differing from what it is not and deferring the trace of what it was and leaning toward those units it will differ from in the future) that constitutes the linguistic system itself. In short, *différance* is the condition of semiological difference and calls into question a privileged self-present, conscious, speaking subject, who is inscribed in semiological *différance* and does not exist independently. The privileged connection between consciousness and meaning-giving is "the ether of metaphysics" (Derrida 1973: 147). Deconstruction shows presence to be an effect of the system in the same spirit that Baudrillard argued use value was an effect of exchange value and the signified/referent were effects of the signifier. Derrida adopts the terminology of differing and deferral—the one side of an opposition is the other deferred (detour for Derrida but alibi for Baudrillard), in its difference from the other (spacing for Derrida and veiled discrimination or unequal weighting for Baudrillard). Symbolic exchange and *différance* are subversive of difference, whether the realm in question is semiolinguistic or philosophical. Derrida, in knowing there is no word for the little "a," affirms it with a Nietzschean laugh, but Baudrillard is dead serious, and he thinks he has anthropological sources for his unalienated, subversive, and positively present symbolic exchangers who swerve around any dialectical negativity (Lyotard 1993: 108–9).

Telemorphic Mauss

In the mid-1970s, Baudrillard "turn[ed] Mauss against Mauss" (1993a: 1). He applied what Gerry Coulter (2012: 51) has referred to as his "one great thought" of reversibility: "the idea that all systems lead to their own demise," a deconstructive principle by another name. Accordingly, the gift reverses into a counter-gift which is a synonym for reversibility. I have written elsewhere about Baudrillard's deployment of the form of death as a virulent counter-gift and I will not rehearse those ideas here (Genosko 1998). This is the first way that Baudrillard reads Mauss and it heavily invests in the "return" portion of gift exchange, relying on the "counter-gift" of incessantly returning in kind or with interest all the apparently unilateral gifts of the social beyond wages and security and the socialist sentiments upon which Mauss insisted. The symbolic is a defiant trap: oblige "the system of power" to receive something it does not want, a counter-gift, which precipitates its collapse. Baudrillard initiates a species of reciprocity that does not rely on balance, but forces the issue of ongoing obligatory exchange where otherwise there would be none. Here, reciprocity does not assist in normalizing and stabilizing social relations. It aims at the very least to destroy the structural theory of value!

I will focus on another, later way that Baudrillard finds to read Mauss. He does so not by writing about symbolic exchange and deploying its destructive capacity, but by writing in a symbolically charged mode of theorizing that challenges one of Mauss's most extraordinary claims: gift exchange is a total social fact. But he does so neither by claiming it to be less or more so but, instead, approaches it in a completely disruptive way, and self-generates symbolic force through a kind of absurd excess. Baudrillard converts the agonistic dimension of the symbolic into theory-writing, and this marks a decisive shift in his later work toward a deeply cynical and irredeemable view of the world, with fewer academic trappings and hence a rather thin web of citations.

Baudrillard's essay "Telemorphosis" contains a provocation that is among the richest in his oeuvre: reality TV is our total social fact. Mauss's upgrading of social facts remained highly suggestive but not rigorous since he alluded to gift exchanges as total through the enfolding, crossing, and manifesting of all the major parts of a social system, either simultaneously or asynchronously. Grasping this total social fact is another matter altogether, for while it requires an encyclopedic knowledge Mauss is careful to diminish expectations through his essay that he will pull all the threads (for instance, he leaves certain tasks for others; he cites the provisionality of his treatments; sometimes he uses "total" as a way to avoid reducing something to a legal or economic basis). This version is on the face of it a sociological outrage and bastardization of Mauss, to whom Baudrillard owes so much. Still it makes one wonder: has television, as Baudrillard claimed, in producing a global nonevent that elevates parody and banality and the farce of the social succeeded better than any radical critique or Situationist driftwork? Where Mauss saw the convergence of all the elements of indigenous societies in gift exchange as total social fact, Baudrillard speculates it is reality TV that gets the different parts of our society working together and permeates them. Baudrillard never really explains why reality TV is this kind of phenomenon on par with gift exchange; he merely asserts that total telemorphosis is evident in reality TV. It would be easier to claim that reality TV is another thread that is woven into the total social fabric of the transmutation of reality. However, he is quite certain that reality TV is much more than a single thread. In taking his chosen route he ruthlessly exploits Mauss's famous undertheorizing of the total social fact, which seems to be in some respect a gauntlet that Mauss threw down for all those scholars who would not dream of a lofty totalizing impulse in the shadow of the master's great learnedness. Likewise, Baudrillard harbors a similar ambition, and doesn't eschew the notion of a totality that is metaphorically interwoven with all of the social's threads, although it is a dead power, in other words, a "total disillusion" that we call simulation: "Today, reality massively transfuses itself into the screen in order to become disembodied. Nothing any longer separates them. The osmosis, the telemorphosis, is total" (Baudrillard 2011: 49).

In his earlier work on simulation, Baudrillard took his readers back to an era prior to the current flood of reality programming with regard to a piece of cinema verité adapted to the little screen, *An American Family* (1973), produced by Craig Gilbert for National Educational Television and distributed by member stations of the Public Broadcasting Service in the United States. This was a twelve-episode serial documentary study of the Loud family of Santa Barbara, California. By the mid-1970s, Baudrillard had already exposed the paradoxes of "TV verité" that *An American Family* ushered in. The producer, Baudrillard claimed, waxed absurd: "'They [the Louds] lived as if we weren't there', which also meant they lived 'as if you were there'– a message to the 20 million or so viewers who followed the series. It is difficult to ascertain the truth of the Loud family: does it belong to the family or to TV?" (Baudrillard 1983: 52) Neither, it turns out, for the despotic gaze of the camera has been displaced, diffused, as the many watch the few; gone are the imperatives to submit to the controlling gaze—you are the gaze, you are the event: "we are all Louds, doomed not to invasion, to pressure, to violence and to blackmail by the media and their models, but to their induction, to infiltration, to their illegible violence" (Baudrillard 1983: 55). The truth of the Loud family is for Baudrillard indecipherable, yet post-panoptic. Some readers of Baudrillard, for instance Grace (2000: 98), have remarked upon these comments and related them to Australian versions of reality programming such as *Sylvania Waters* that equally make Baudrillard's point about the implosion of the distinct poles between viewer and viewed, question and answer, referendum and referent; even the medium itself has collapsed into its messages and as a communication process the poles of sender and receiver are henceforth untenable.

Much later, ruminating on the French Channel M6 reality show *Loft Story*, Baudrillard emphasizes that at the heart of reality TV is an uninteresting, nonoriginal event, whose power is to generate the fascination, first, of audiences, and, second, of critics. Baudrillard underlines, however, that this nullity and banality is powerful, even if it only allows for a differential viewing experience: the viewer is always slightly less idiotic than the reality TV program on the screen. The question that holds Baudrillard's attention concerns the "experimental niches" of reality TV situations—apartments, islands, and other microsituations. Are these enclosures, little theaters, cut off and isolated in some manner, or do they jump their experimental status as "universal metaphors" of the osmosis, what he calls the "telemorphosis" of the world? "Nothing any longer separates" the screen and the world. Telemorphosis of the real traps everyone. We are all extras on the next call for contestants; we are on both sides of the screen, playing our parts, ready for sudden notoriety for no good reason. We are all Trumans! For Baudrillard reality TV announces the end of merit: there is no need to earn fifteen minutes of fame. It will be granted or taken, as the difference doesn't seem to matter.

According to Jeffrey Ruoff in his study of *An American Family*, in addition to factors relating to program conception and filmmaking, editing mitigated reality: "the producers discovered that watching footage shot in real time was strangely unlike real life" (2001: 40). It was, in a word, boring. But boredom was also an existing aesthetic inherited from Andy Warhol. Moreover, it was also fuzzy, as one editor

David Hanser recalls the experience of viewing hours of film of Loud son Lance and then seeing him in person: "Reality began to get blurred." Here is a sense of the vertigo of implosion à la Baudrillard and a fall into a fuzzy, speculative universe. Ruoff does not work out the implications of this boredom in relation to what TV critics, for instance, point out about reality programs such as *Big Brother*: the boring factor as legitimation. But for Baudrillard there is no such evidence, no such recourse to proof: there is only an accelerating "banalization of the world" that reality TV brings to a crescendo: reality is uploaded to the screen. "Running gags" between reality and the screen no longer works and is relegated to the cheapest form of comedy—the unfunny kind.

Ruoff is incisive: *An American Family* simply lacked "historical context." This grounding didn't, he explains, survive the proposal stage: "Craig Gilbert did not make the series he described in his NET proposal," which was full of references to the mediascape of this family's life as well as familial histories. In the end what won out was the Louds' separation. The absence of historical context ("reality") was made a topic in its own right: in the end, "Gilbert was satisfied to portray a family not directly connected to the issues of the day in order to claim that Americans were alienated from politics and active citizenship" (Ruoff 2001: 23). "Reality" was chased from the stage for the sake of the serialized drama of separation and divorce; reality was easier to bear when its absence could become a symptom of disenfranchisement. This was not exactly telemorphosis proper, but a preparatory state when representation was still a stake.

Baudrillard has roughed-in the trajectory of reality television from the 1970s to the early 2000s. His is not an exhaustive accounting, but a landmarking. For him reality TV is a privileged funnel for the telemorphosis of the real, and this means television is his choice medium for such a transfusional absorption. After telemorphosis, is it still possible to speak of the screen? Indeed, is any reality TV program worth mentioning by name (especially *Big Brother*)? What are the differences between them, anyway?

Perhaps the question that Derrida posed of Mauss should be put to Baudrillard: his discourse on *Loft Story* in a way arrests the process of telemorphosis by preventing it from escaping the confines of the program. The gift cannot make itself present, and for Derrida in *Given Time*, Mauss speaks of everything *but* the gift (Derrida 1992: 24). The gift's condition is that of Being: disclosure and concealment. And as far as Derrida is concerned, Mauss qualifies "total" as something other than itself, making the perception of the whole impossible. How does this apply to Baudrillard? The process of telemorphosis stalls if the medium into which reality is transfused is still identifiable and describable after the process is well advanced. It must be possible to radically forget the screen, its features, frames, and textures. It must be possible to no longer see beyond the screen in the state of full telemorphosis. Derrida displaces the total (except the gift); Baudrillard claims that telemorphosis is total (with no remainders). It is of no use to compare the little to the big screen; cathode ray to flat screen; neither to find where precisely the screen begins and ends. And in the end the program must evaporate as a stable point of reference. So, too, must reality TV as a distinct genre. There is only a telemorphosed reality: full telereality: "Today, the screen is no longer the television screen; it is the screen of reality itself" (2011: 49–50). In the end, Baudrillard, in the spirit of Derrida, displaces reality TV's and television's roles as totalities: telereality is neither restricted to one genre nor one medium.

Liquidation of signification

In his later writings on power, Baudrillard adopts a distinction between domination and hegemony: the former is a dual relation of asymmetrical force, and the latter an integral, networked, calculated, cybernetic relationality. This distinction, while in some important respects consonant with Deleuze's observation that we have passed from disciplinary to control societies, is marked by simulation and its effects. Both Deleuze and Baudrillard, it is worth noting if only in passing, consider modulation—for Baudrillard by means of models (1993a: 56), and for Deleuze a continuously undulating form of self-guided molding in a diffuse space—as a hallmark of the society of hegemony/control. And it was undoubtedly Baudrillard, and not Deleuze, who announced the eclipse of panoptic perspective by means of the transition from surveillance to deterrence (1994: 29). However, I want to focus on Baudrillard's return to what I will call a retheatricalization of simulation.

The contemporary form of simulation in our society of hegemony is masquerade:

> Hegemony works through general masquerade, it relies on the excessive use of every sign and obscenity, the way it mocks its own values, and challenges the rest of the world by its cynicism. (2010: 35)

Masquerade, Baudrillard specifies, is an event involving the liquidation of signification by means of signs, such signs proliferating in a parody of referentiality: "the profusion of signs parodies a by now unobtainable reality" (2010: 35). The implication for domination as for power is that it cannot escape being engulfed by such profusion, hence, "power is only the parody of the signs of power—just as war is only the parody of signs of war, including technology. Masquerade of war, masquerade of power" (2010: 35). Engulfed in a parodic world of semiorrhea, meaning itself is abolished as apparently any recourse to positive values becomes mired in a self-liquidation or a self-emptying like "signs emptied of their substance" (2010: 36). Baudrillard shows no interest in post-signification or a-signification as positive possibilities. Hence, "masquerade of war, masquerade of power" (2010: 35). But can one equally try to claim: a masquerade reverting to its opposite, an unmasked face or a face more virtual than a mask?

All of this is reminiscent of Baudrillard's characterization of the first stage of simulation (counterfeit) as a theatrical illusion in *Symbolic Exchange and Death* and elsewhere. Forgery, deception, stucco, theatrical scenery: every anxiety-inducing "simulation effect" will eventually disappear with the second order as it is crushed by the machine and industrial labor, by robots, serial production, and dull repetition. The "end of theater" is marked by a transition from the automaton to the robot: in the loss of play, all appearances are absorbed into a system without semblances and dissemblances (1993a: 54). Perhaps it is clearer to put this issue semiotically: the counterfeit sign "dreams," Baudrillard (1993a: 51) suggests, of the sureness of its capacity to signify and refer that it has lost. But where obligation was, reason is now found, and the sign, "free, emancipated, and democratic," is released from its hitherto cruel restrictiveness within a clan-based social system. Stucco and other "artificial signs" not only take advantage of the space of illusion and imitation in order to play with the differences and question the relations between counterfeit and real, but are also implicated in social relations and power. Stucco is for Baudrillard the figure of a single, universal substance (not merely material but with representative immateriality) capable of imitating everything and entering into unlimited numbers of combinations, and which theatricalizes the social ambition of the bourgeoisie. Or not quite. The counterfeit is still anchored in substance and not in social relations and structures, but it points in the latter direction by means of its attributes: general equivalence, indestructibility, control, and pacification through "political and mental hegemony, the phantasy of a closed mental substance like the Baroque stucco angels whose wing-tips touch in a curved mirror" (1993a: 53).

With masquerade, one might expect that not only theater but parody return, no longer left behind in an earlier order of simulation by Baudrillard in advancing toward the hyperreal and the erasure of the distinction between the real and the imaginary. What, then, is the role of (re)theatricalization when the real is no longer intact? Moreover, what becomes of the optimism (1994: 121) of theatrical illusion? Hegemony is on the contrary saturated with arrogance, cynicism, and hypocrisy (2010: 38). Any hope of critical negativity, subversion, and transgression has been dashed. Hegemony works by means of a blanket proliferation that is parodic: "total masquerade." Indeed, Baudrillard also aligns masquerade with mockery. Masquerade's mockery of the liquidation of the real, value, and representation is for Baudrillard the "principle of government" in the era of hegemony:

> The new strategy—and it truly is a mutation—is the self-immolation of value, of every system of value, unhinged by simulation of self-denial, indifferentiation, rejection and nullity as the triumphant command. (2010: 50)

Baudrillard's interpretation of the Watergate "scandal" is a foundation for the application of his critique to contemporary events, namely, the 25th anniversary of the Tiananmen Square massacre. Baudrillard's diagnosis of the "vicious curvature of [...] political space" (1994: 18), like the wing-tips of the aforementioned stucco angel, but this time clad in mirrors, reveals how claims about scandals annul themselves and revert to their opposites. The idea that value self-immolates is very distant from the demarcated role of symbolic exchange. Scandal is a masquerade suited to a morality play in which it is impossible to choose between empty ideologues and their critics. The only scandal here is the effect of scandal itself in a political field completely "unhinged by simulation."

If, then, a Baudrillardian position on events cannot be held because it does not exist outside of the field it describes, namely, the field in which the indistinction between poles and positions is impossible to determine, how are we to grasp his claims about hegemony and regain his conception of masquerade? Masquerade is simply Baudrillard's way of underlining the cancellation of differences and the inability to isolate specific processes that rely upon some principle that would save them from indeterminacy. Masquerade is not joyous, not celebratory. It is still, as Deleuzians might put it, indexed to degradation and loss, as opposed to the production of differentiation and the eternal return. Baudrillard remains on this view too Platonist, and too nihilistic. Yet he insists that his brand of nihilism has its passions such as fascination and melancholy. But his confession is in doubt because in itself "nihilism is impossible" (1994: 161), that is, too "determined" and too stable. Therefore, he pushes past nihilism into a more agonizing condition in which power, proliferating with a self-focused virulence, cannibalizes itself toward self-destruction (2010: 61–2). And in the process, mourning becomes an appropriate affect since all of us would be implicated in the performance of an "unimaginable adventure" (2010: 81) of self-obsolescence and capitulation to the machinic order which automates our irrelevance and ensures our disappearance.

Baudrillard's lessons for security are stark. Once the edifice of representation is jettisoned for that of simulation, any recourse to a preexisting point of measure is also lost. Within this order, Baudrillard himself has appeared to augment its intensity. Think of the development from processed war to processed world from the period of the Gulf War to his *Agony of Power* book. The war that will not have taken place played with the lingering critical notion that war has a theater and an event timeline; provocatively, it could find neither. Bill Bogard's (1996) lessons on this point are worth considering anew. Bogard imagines that simulation is fully achieved when it can neither be delivered through any identifiable medium nor framed by a technological device: zeros and ones for as far as anyone knows. The surface is itself not perceptible. Transparency is given in advance and without windows. The medium falls away. In this way, any effort to (re)capture a reality either from an idealist or realist position must itself be simulacral. Extrapolating from this, Baudrillard insists that the spatial and perceptible conditions no longer exist in which traditional relations of "force and violence" can be fought (2010: 93). Hegemony is, then, not only post-political, but post-human, entailing obsolescence of human being in a "non-Euclidean space" of unpredictable events (2010: 89).

Whither masquerade? For Baudrillard, masquerade is a concept that retains nothing of the play of reality and its simulation once rehearsed in the hegemonic order. It is only awkwardly indexed to specific reproductive media and to material platforms that require referential stability. Today, masquerade suggests the affect of global panic, of pseudo-events like bird flu scares and Y2K bugs. This also includes the euphoria produced by a "silent conjuration": "our secret [imbecilic] delight in seeing the stupidity and corruption of those in power" [like Toronto's former Mayor, Rob Ford] (2010: 46). But this is not new as Baudrillard explained much earlier in his analysis of the French Socialist Party, *The Divine Left*, when he included entertainers turned politicians like Yves Montand and Ronald Reagan as examples of "the revenge of the people against the political class" (2014: 107).

Finally, Baudrillard's later position calls into question a critical recourse to writing—with neither a medium nor a message! Writing is displaced onto splogs and content farms (Brunton 2013: 159–63) and his earlier belief in poetic dispersion, pataphysical love of paralogy, New York graffiti, and the delicacy of passwords, anagrams, and antitotalizing and antiteleological fragments can longer find a place in a world of networks and machine-to-machine communication: vulgar, referentless, programmable signals. Does it matter if the force of the symbolic survives in theory-writing? After all, what is symbolic exchange once signs have set themselves ablaze and communication takes place beyond the code? This is where the gift returns with a terrible vengeance. For one accrues prestige by means of unlimited expenditure, by reckless destruction of one's property, in a masquerade of indifference to one's possessions, that is, a feigned interest in self-gratification, but mixed with knowledge of the creation of hardship and likely self-harm. Baudrillard's anthropological imaginary returns in the end to Bataille's valorization of expenditure, against rational and utilitarian usage of goods, as self-fulfillment of the things lost. This is the non-co-optable exuberance of which Baudrillard, too, dreams.

The politics of response

Baudrillard's meditations on the meaning of response in a society of nonresponse are tightly bound up with his theories of media and communication. Both are examples of unidirectional gifts without real returns; they reject responses of any kind in favor of integrated "feedback" and facile solicitations, while communication is without a genuine co-relationship.

"Power belongs to the one who can give and cannot be repaid," writes Baudrillard (1981: 170), pointing to a profound disequilibrium in social relations and the incapacitation of repaying, or returning in kind or with interest. A counter-gift is thus excluded, but if it can be reinstituted, as a possibility, it would disrupt power and threaten its monopoly. The return of symbolic exchange takes many forms in Baudrillard's thought, from the immediate, reciprocal, and antagonistic exchanges on the streets during protests—these are "beyond the code" as he puts it, citing graffiti against advertising; handmade posters and buttons against official insignias and copyrighted logos; witticism and neologism against dominant discursive formations. The strength of these kinds of responses may be seen in their destruction of media, as they are imagined to be medium-less, without intermediation of any kind. For Baudrillard, this is a deconstructive task that advances from the separation of the signifier and signified to the sender and receiver: all separations and their simulacral resolutions must be destroyed. There is more Bataille here than Mauss: regaining symbolic exchange against the degradations of simulation overcomes artificial separations and reinstates genuine communication among intimates without the estrangements of signs.

3

Delineating Info-commodities in the Age of Semiocapitalism

This chapter explores the theory of semiocapitalism or semiotic capital most strongly associated today with two strands of Italian post-autonomist thought represented by Franco Berardi and Maurizio Lazzarato, whose work on Guattari was discussed in the first chapter. However, Guattari's original insight into the mutation of capital was to describe it in terms of procedures of semioticization that went beyond monetary and financial modes by plugging into an emerging global and machinic network. Guattari focused his attention on minimization, precarization, and networked production and the implications of these and other factors for subjectivation while developing an overall new type or phase of what he called Integrated World Capitalism (IWC): a theory of globalization before the letter developed collaboratively in the 1980s with French philosopher Eric Alliez and characterized as a stage of postindustrial capitalism marked threefold by modes of machinic production and their integration (i.e., by means of permanent crisis); by dominant semiotic-economic systems in which the market becomes transnational (i.e., global capital before the letter); and a state that becomes minimal and speculative (i.e., outsourcing public services). IWC is akin to postindustrial capitalism and hints at the emergence of the importance of immaterial labor in which human semiosis (general intellect) is directly and immediately productive of value with minimal material mediation (i.e., in the online activities of social media websites which extract and exploit user data in the form of search terms, posts, likes, links, friends, groups, etc.).

For Guattari writing in *The Three Ecologies*, IWC organizes itself for the production of "signs, syntax and … subjectivity" (2008: 47) and in this respect is the immediate precursor to semiocapitalism in which capitalist production has become semiotic and "seizes individuals from the inside" (1996b: 220) so that capital looks inside and expands intensively, becomes real and not only formal (capital looks outside and incorporates noncapitalistic processes). As Antonio Negri (2008a: 25) explains in *The Porcelain Workshop*, there is no outside, nothing transcendent, as everything is integrated: "nature and humanity have been transformed by capital. From now on, all aspirations to alterity … are not only outdated but also vain. And yet: from inside this fetishistic world, the antagonism of living work is affirming itself and resistance is building." It is not difficult to find the political foundation of Guattari and Negri's friendship and theoretical collaboration in Guattari's corroborating affirmation of resistance within integrated capitalism: "it does not seem at all absurd to anticipate that the development of new collective responses … coming from the greatest variety of horizons, might finally succeed in bringing it down" (1996a: 246). He notes popular molecular disturbances and struggles around the globe in the 1970s and 1980s (Sandinista National Liberation Front/Nicaragua, Solidarity/Poland, Autonomia/Italy, Workers' Party/Brazil) as evidence of resistance to the unidimensional subjectivities (serial and standard) and social segmentations (precarity of labor, reified generations, dismantled class alliances) of semiotic capital.

These initial orienting considerations form the basis for Berardi's explicit theorization of semiocommodities as the coagulation of capital flux in semiotic artifacts that do not immediately accede to hard materialization; this process of slowing down and pooling describes the immaterial factors of existence betwixt immateriality and materiality. Berardi extends Guattari's insights into the machinic time of capital as a semiotic operator and this theory dovetails with Lazzarato's hypothesis of immaterial labor which underlines the significance of labor's intellectualization and immaterialization without completely eschewing either embodiment (for the sake of purely cognitive labor) or materiality of its products (in favor of intangible forces like affects). Rather, as I will show, the theory of semiocapital rises to the occasion by explaining how immaterial factors find soft semiotic substantializations prior to and, indeed, involving the deferral of any kind of materialization. From a formal semiotic perspective, the key background relation this chapter concentrates on is between semiotic and material strata. The lingering dichotomy between abstraction and instantiation, between language and speech, which I broached with reference to their flushness in Chapter 1, forms the conceptual context of this investigation.

The concept of semiocapitalism aims, first and foremost, at vanquishing false antinomies and updating stale distinctions like those belonging to structuralism and Marxism, that is, between the signifier and the real, and between desire and the economic infrastructure, respectively. Specifically, the concept has the ambition of capturing in all of its complexity the relations, spatially interactive and temporally complex, between the semiotic and material strata in the context in which informational commodities are created by info- or digital laborers, described as immaterial labor or semiotic production of knowledge workers. Immaterial labor is a rich semiotic concept that refers to both labor processes—increasingly informatic and communicational—and the provision of the cultural content of products, requiring capacities of aesthetic judgment, taste, and style sensibilities not normally associated with work and traditionally the province of the bourgeoisie and children of privilege (Lazzarato 1996: 133–4). An info-product like a piece of software that contains a world like a video game has a corporeal manifestation in an autonomous object, a plastic cartridge, dvd disk, or similar device that plugs into a hardware system that may be networked; affective production (care giving in the health sector; or greeting customers in the retail and entertainment sectors) that takes place across the vast service sector often has both manual and intellectual aspects (Hardt 2006: 48–9). In short, the semiotic subtlety of immateriality is suggested in the complementary statements that its products are not entirely immaterial—and immaterial labor is not entirely cognitive or affective, either. This does warrant a tentative conclusion that the primacy of the material thing is put in doubt in some manner (perhaps "outweighed" by the "less tangible," Dyer-Witheford and De Peuter 2009: 4–5), and a reticence to completely commit to the role of cognition in the immaterial labor hypothesis, especially in the service sector. However, semiocapitalism is an invitation to explore the semiotic–material relation through the lens of labor as both intellectual and affective but also marked by requirements of participation and collective cooperation (work in teams and groups, open office spaces, constantly communicating within distributed networks): relaying information is non-negotiable.

Critical responses to the immaterial labor hypothesis have ranged from the preposterous—"it is transparently absurd" (Graeber 2008: 6)—to an acute analysis of what the polemics against it exclude, namely, the failure to appreciate its hybridity and how it draws material labor within itself (Negri 2008b: 69–70). Whether immaterial labor is grasped tendentiously as a tendency—"borne out by certain technologies (financial, media, digital) and subjectivities, which, whilst not amounting to a quantitative majority, is in some sense the hegemonic cutting edge of capital" (Toscano 2007: 84–5) or, any combination of the less (tangible, about things, and less important, like the immediate application of labor time [Dyer-Witheford and De Peuter 2009: 4–5])—is not a moot point but one that would wisely seek a balance with positive assessments of what immaterial production or, better, "being" is more and above all about: intellection, affects, relationships, cooperation, language, communication (Negri 2008c: 214). It is not my goal to repudiate the dislike of a labor so "untenably diluted" (Coté and Pybus 2007) that it needs supports; nor to deny that a labor that results in no material goods and thus points to the generalization of the predicament of women's domestic labor whose immaterial goods are often inseparable from their production (Fortunati 2007) seems to offer no more than another confirmation of the latter's subordination. Indeed, over almost ten years from the publication of *Empire* to *Commonwealth*, Michael Hardt and Antonio Negri's presentation of the significance of immaterial labor has accumulated a number of models: where first the intangible products of immaterial labor were modeled on affective (health and entertainment sectors) and symbolic-analytic services (computer and high-tech sectors) (2000: 290–1), the most recent work more fully acknowledges the feminization of labor without gender equality as a model for how affective labor, which is "required of women disproportionately on and off the job" (2009: 133), has come to assume a central place for capitalist valorization. Today, Berardi (2015: 90) sometimes uses semiocapitalism as a synonym for financial capitalism: "Financial abstraction is only the extreme manifestation of the predominance of semiosis in comparison with physical production."

Acknowledging that immaterial labor processes are not entirely cognitive but may involve manual applications, and that the products of such labor may have material manifestations in autonomous objects, suffices to affirm the kind of hybridity insisted upon above, and cannot be read as mere capitulation to the necessity of corporeal/material mediation. Together Hardt and Negri (2009: 132–3) insist that immaterial factors may "outweigh" the material aspects of commodities, and the labor that produces them is both corporeal and cognitive. However, if immaterial labor points to a fundamental change within post-Fordist production in which human semiosis ("general intellect") has become directly and immediately productive of value without any material mediation, even while remaining embodied in some manner (Virno 2001), how is this immateriality maintained while acknowledging hybridity? In order to grasp the semiotic dimension of contemporary capitalism, then, a hybrid theory of semiosis is required.

The neologism "semiocapitalism" combines a general semiotic and a contemporary formula of capitalism—which may or may not be the highest—and also participates in a periodization of sorts since the concept references the flexibilities of post-Fordism, evoking mobile productive spaces (postfactory), the rise of precarious labor for whom life is indistinguishable from work, and the financialization of the economy. Most importantly, a fundamental shift has occurred in post-Fordist production. Consider the implications of Berardi's (2009: 21) remark: "The rise of post-Fordist modes of production, which I will call Semiocapitalism, takes the mind, language and creativity as its primary tools for the production of value." In other words, the production of value in post-Fordist capitalism occurs through the appropriation of general human semio-communicative capacities, called by Marx in the *Grundrisse* (1972b: 706) "general intellect," in a manner that is lightly if at all mediated by things or product-objects. General intellect is often referred to a "general knowing." For example, Paolo Virno explains: "We live in a time, the post-Fordist era, in which human nature has become an economic stake. Every aspect of human nature ... constitutes raw material for production" (2006: 24). Human semiosis itself is directly and immediately productive of value through its combination in semiocommodities. However, how light or indeed absent this mediation may be—and some such as Akseli Virtanen (2004: 226) have insisted it takes place "without the necessity of mediation or a corporeal form"—remains a key point for consideration. Virno (2001: 1) was one of the first to advance the view that all of the traits of general intellect, including "formal and informal knowledge, imagination, ethical tendencies, mentalities, and 'language games' ... function themselves as productive 'machines' in contemporary labour and do not need to take on a mechanical body or an electronic soul." This is a refined version of immateriality. Another version takes a more applied approach and uses the competence required to enter a search term into Google as an example of the transformation of common intellect into network value through the PageRank algorithm (Pasquinelli 2009); clearly, in the background of this example is an embodied user. A further critical view is that of Gerald Raunig who, in *Factories of Knowledge*, questions the generality of intellect, not as more machinic and more living than ever before, but whether or not it is "sublated into a universal unity" (2013: 65). If this is the case, it restricts our understanding both of the singularities and their relations within general knowledge by falling back on pregiven and preformed knowledge. For Raunig, this necessitates the invention of a "transversal intellect" that "emerges in traversing the singularities of thinking, speaking, fabricating knowledge" (2013: 66). Thus, the need to acknowledge the potentialities of living labor for new emergent formations (i.e., general assemblies of Occupy Wall Street and so-called organizationless organizations).

Berardi presents in *The Uprising* a very different version of the fate of general intellect by describing how its potency ("infinitely productive"; 2012: 77) fueled the euphoric dot-com years. The disorientation of drug-fueled semiocapitalism resulted in depression and disempowerment for cognitive laborers as psychic energy was drained away in a dependency on unsustainable speed and boundless growth. This leaves general intellect looking for a body, for new ways to thwart capital, and Berardi points to WikiLeaks as a prime example of its latest innovative autonomization. This example raises more questions than it answers, however, as it is obvious that Julian Assange has been isolated in a place (Embassy of Ecuador, London) from which he cannot, at the time of this writing, move yet. WikiLeaks is itself part of the abstract network of the Internet, if one doesn't dwell too much on the undisclosed locations of its servers.

Lazzarato asks his readers to cast off their "factoryist prejudices" in order to grasp the temporality of the cycle of production that is summoned by the capitalist for the duration of a task after which it disappears back into the networks that made it possible. Semiocapitalism seemingly carries us far from simplified industrial labor, the traditional workday, and the factory floor to the complex tasks undertaken at any time, seemingly any place, by cognitive laborers according to their singular subjectivities, all yoked to increasingly mobile and extensive electronic networks of exploitation, as well as those professionals who appropriate the cognitive surplus value of designers, philosophers, engineers, computer scientists. All of life, what may be called life-skills, and experiences can be appropriated by capitalism. Yet in this for Berardi (2015: 49) "people are erased from the space of work" and nonlocalizable labor "can be transferred, fragmented, fractured and finally recombined in the abstract space of the Internet."

While the concept of semiocapitalism might be taken as a variation on the infiltration of information into the process of production (Guattari 1996a: 244), hence info-capitalism and its globalization as integrated world capitalism in which older segments of territorialized capital appear less and less relevant, the analysis of how production becomes semiotic entails the theorization of semiotic operations answering questions about which signifying relations are at issue and what are their salient traits; launching critical inquiries into the relations between capitalist power and signification; and describing the limits of the formation of semiotic substance expressed in the semiotic–material relationship. In other words, how do semiocommodities express the connection between semiotic and material fluxes when the thing-commodity has been displaced by productive signs or part-signs working directly on and with the raw material of human intelligence? In teaching how form and matter combine by swerving around substance, Guattari provides an indirect lesson on how semiocapital approaches and orbits around substance. Instead of specifically forestalling the language-specific linguistic formation of matter, the question becomes how to describe soft, partial, semiotic substantializations of immaterial forces like affect?

It was Guattari who founded a conceptual language adequate to the investigation of semiocapitalism. His approaches and concepts have inspired and informed contemporary theorists like Lazzarato and Berardi, among others, whose work has opened up a troubling new chapter of capital's misadventures. From a semiotic perspective, semiocapitalism engages what might be considered a classical semiotic problematic: the description of the formation of matter given most explicitly and influentially by Hjelmslev. He thought of unformed matter (purport) as sand, an "amorphous [thought] mass" fundamentally receptive to form understood as the molds of different languages (1969: 52); he figured the projection of form onto matter as the shadow of a net, a grid, cast onto the undivided sand: a form cast onto an amorphous material that yields a substance seen in the shadow. The move to figure form as an abstraction, yet a constant that is necessary for the variable substance, inspired Guattari to recast form as an abstract machine irreducible to language—ensuring its abstract purity (like a lexical gap without substance). It is not the net but its shadow; it is not the form but its function, hence an unformed function. Guattari does not restrict his use of "unformed." In Guattari's thought, an abstract machine is a function rather than a form: "functions are not yet 'semiotically' formed, and matters are not yet 'physically' formed" (Deleuze and Guattari 1987: 141). The abstract machine is a "pure Matter-Function" that does not form substances and lacks distinct strata of its own (i.e., expression/content planes). The conceptual language that Guattari adapted from Hjelmslev is additionally loaded with his adaptation of the diagram from Peirce: the abstract machine possesses only minimal traits (particles) and intensities (tensors) and may be characterized by its semiotic reticence, especially in the moves to pure function, abstraction, and "not yet" that maintain the independence of the abstract machine. Thus, Guattari cannot be said to advance a semiotic hylomorphism of a fixed form imposing itself on an unformed, amorphous matter, as his deformation of Hjelmslev is precisely designed to avoid this outcome. Like Deleuze, his sense of form and matter transforms into active meetings between forces or force–matter relationships, marked by reversals and reciprocities and presuppositions (Massumi 1992: 11; Smith 1996a: 43). Guattari replaces forms with abstract machines, and they do not consist of substances because form is substance as they lack distinctiveness: "they [the machines] know nothing of form and substance" (Deleuze and Guattari 1987: 511). There is an important condition attached to this: since abstract machines do not control all of the activities of articulation and disarticulation that take place on the strata and escape it, substance is not excluded in the thickenings and *coagulations* that arise (Deleuze and Guattari 1987: 502 and 513). Here, *coagulation* is a kind of consistency on the strata that functions below the threshold of well formedness consisting of loose and irregular aggregates and lingering destratifying forces.

The conceptual indebtedness of semiocapitalism to this reticent and molecular semiotics reveals itself in the way that it is distilled into a function and does not (yet) culminate in a distinct, stratified representation, a physically formed substance; substance, in this account, is "bracketed" (Massumi 1992: 14) in order to release the dynamic elements of encounters. Further, what makes this approach molecular is the desemioticizing effect of forestalling the move to substance and well formedness by maintaining independence from the direct determination of formed substances: "unformed matters and nonformal functions" (Deleuze and Guattari 1987: 511). An abstract machine directly "pilots" matter prior to the strata of pertinent semiotic distinctions between signifier/expression and signified/content. Traits are "not yet" formalized into signs like icons, indices, and symbols with their respective objects. In short, Guattari reinvented the terms and relations of Hjelmslevian theory but not only by making form an abstract machine independent of substance but by bringing unformed matter back into the picture after it was left outside of glossematics for science to study. In so doing, Guattari (1977: 265) broke the "pincers" of proliferating dichotomies at the heart of structuralism, without himself establishing a new dualism between abstract machines and formed semiotic strata. He advanced on Hjelmslev by treating unformed matter as a phylum that bears singularities in its movements and nonformal functionality as a diagram that carries alien expressivity in its movements, such as aberrant poetic syntaxes full of anacolutha—agrammatical constructions.

The theory of semiocapitalism received a major conceptual and semiotic impetus when Guattari's imagination was activated by the provocative idea of function–matter relations outside of signification. Therefore, thinking about the relation between semiotic and material as conjoinable mutual intensities engages the materiality of the semiotic and the semioticity of the material. The terms of this encounter are worked out by Guattari through the substitution of collective for individuating enunciations; nonhuman for human-centered fluxes, and artificial for natural referents. In a number of texts, Guattari had recourse to technological examples to explain the potential he wanted to explore: machinic conjunctions between semiosis and matter might involve either the "piloting" or "guidance" of fluxes of equations ("logic machines") and electrons (1978: 270). Further, he invokes one of his own inventions, *la grille* (the grid), that he used at Clinique de la Borde: an abstract machine written on a piece of poster paper that conjoined the fluxes of time, labor, tasks, and money into a rolling schedule of work assignments that broke from hierarchical organizational prerogatives and existing divisions of labor and specialism and for a time put brooms in the hands of doctors and medicines in the hands of patients and kitchen pots in the hands of groundskeepers and dirty linen in the hands of administrators, and so on (1998: 271). The overarching political goal was to put groups into contact with the fluxes when they were otherwise prevented from doing so by the institution's existing organigram of power.

The limits of structural homology and the critique of formalization

Part-signs, it may be recalled, are akin to signals, although care needs to be taken in not collapsing together the signals of a-semiotic or natural encodings (outside the semiotic strata; pheromones) at the microbiological and chemical levels and signals of automated, machinic infotechnological environments (algorithms).

Extrapolating from Deleuze and Guattari's *Anti-Oedipus* (1977: 68ff), the connective syntheses of production connect part-signs in an automated network of trigger actions by giving start and stop orders (Guattari 1995: 49). Part-signs in a material machinic informatic network are manifest, for example, as start and stop characters and data separators on the track of a credit or debit card's magnetic stripe. Connection gives way to disjunction and antiproduction when a network is compromised by any number of hacks that visually capture PIN numbers, or intercept communications within a credit network or by means of fake hardware placed in an automated banking device. Disjunctive energy using various deceiver codes or devices disjoin connections so that hacking unfixes and deviates digital communication; there is, here, a process of vacuolization at work. While some of these hacks fall back onto deeply banal processes of redundant bombardment (denial-of-service attacks), some schemers try to get rich from their spam, and others just herd as many zombie computers into vast criminal armies or engage in simple juvenile delinquency. The subjectivities that emerge alongside the info-machine as effects of the mingling of production and antiproduction are sometimes Bill Gates and sometimes fifteen-year-olds doing it for the lulz, white hats and black hats, hacktivists, or netbot herders. Some become rich and powerful; others go to jail and write books about it; many just disappear into adulthood, or are turned into informants by the FBI, but perhaps it is closest to the spirit of the conjunctive synthesis to say a little of everything is assembled by each.

It is instructive to first turn to Baudrillard for he explains semiolinguistically how semiocapitalism may be described as an "infinite game of mirrors" (Berardi 2007: 76–7), or play of negative, structural differences; yet his recourse to linguistic signs is only partially compatible with the semiotic theorizing of Guattari. The latter places greater emphasis on Hjelmslev–Peircean hybrids as an antidote to the imbalance exposed by Baudrillard, among other objections, in seeking nonlinguistic "springs of creativity" (Guattari 1996a: 146) against the tendency of linguistic overcoding to be readily recuperable by and in collusion with capitalist power. For Guattari (1996a: 144), "capitalist powers never cease 'rethinking' in detail each semiological 'allocation' [beginning with children's language training]."

The two-sided commodity was analyzed as a familiar form in Chapter 2. Baudrillard thought he could expose the ideological imbalances lurking in what appeared to be structural correspondences—for instance, the homology between the commodity and the linguistic sign. Exchange value and the signifier have a strategic value greater than the tactical value of use value and the signified. Binary oppositive structuration is never symmetrical since each antecedent term produces its own "alibi" as its consequent term. Use value and the signified are effects or simulation models of their antecedent terms. They are produced respectively by Marx's analysis of commodity fetishism in terms of exchange value, while semiolinguistics privileges the signifier as its principle of circulation and regulated interplay. Hence, there is a fetishism in the signifier's capacity to provide shelter from a real that is nevertheless prolifically signified. This is the precise meaning of the structural revolution of value: real referentiality is annihilated and simulation of the real wins out.

The consistency between Baudrillard and Guattari is that power invades the sign-form in a number of ways: by privileging one restricted element that creates dependency, for example, an "invariable" like form in Hjelmslev. Or the signifier that controls the signified—the former has a despotic ambition to "flatten" the latter (Guattari 1977: 256); any possible relations with intensive realities are controlled by highly artificial constraints—like referentiality—or double formalizations, like syntagmatic and paradigmatic axes; by placing limits on articulations—"double" as opposed to many; and producing an impotent subject who ignores social and political processes that influence expression and content, and the uses of meaning, trapped in an inward-looking circuit and beholden to what is inscribed in representation (a cognitarian subject of the statement, or semiocapitalist subject of enunciation), "rectifying" every nuance and flux by overcoding them in doublets of formalization. For Guattari, signification is inseparable from power: "all stratifications of power produce and impose significations" (1977: 306). Paraphrasing Guattari, dominant significations—values, roles, small and large identifications from sex to nation, rules of conduct, translations between codes—that are formalized in social systems are conjoined with an automated a-signifying semiotic machine that is itself without any meaning (it produces machinic rather than signifying redundancies) but whose task it is to produce the outcomes desired by the system at issue (i.e., the "proper" contents/signifieds). As Julia Kristeva has observed, semiotics homogenizes heterogeneity: it "loses hold of it" (1986: 30). Guattari observed that capitalism relies upon and manipulates a-signifying semiotic machines—these are the so-called wheels of the stock market whose ups and downs do not have an essential connection with meaningful symbolic or semiological systems in reinforcing economic stratification through the secretions of affect—the Dow-Jones had a good day!

Both Baudrillard and Guattari argue, to quite different ends, for the need to bypass mediating terms in the relationship between semiotic and material fluxes by criticizing the sign's political economy. The important point is that this move clears the way for creative thought and experimentation—for instance, Guattari's theorization of the part-signs of a-signifying semiotics that does not require semiologically signifying, subjectifying (individuated), representational mediation, though it may eventually have recourse to such substances and meanings (other kinds of signification). For Baudrillard, semiological detours are not required for immediate symbolic communion.

Capital is a semiotic operator

With this unique semio-material machinism in mind, I want to consider the changes that semiocapitalism evokes. Capitalism becomes less interested in an older form of standardized physico-social labor defined quantitatively ("value can no longer be defined in terms of average necessary work-time for labour," Berardi 2012: 74) and more interested in a nonlocalizable qualitative process by means of which surplus value may be generated but without work in an industrial sense of the term. Capitalism incorporates into this process "every semiotic system" (Deleuze and Guattari 1987: 492). That is to say, all of the social enters production. "Capitalism," as Guattari states, "seizes individuals from the inside" (1996b: 220). This formidable production of subjectivity has led Lazzarato (2006b: 3) to reject the description of contemporary capitalism as a mode of production for the sake of a "machine of subjectivation" which the process of immaterial labor brings to a pinnacle: "On the one hand, the individual brings the subjectivation process to its pinnacle, because in all these activities [of self-entrepreneurship] s/he involves the 'immaterial' and 'cognitive' resources of her/his 'self'; while on the other s/he inclines towards identification, subjectivation and exploitation, given that s/he is both her/his own master and slave, a capitalist and a proletarian, the subject of enunciation and the subject of the statement." Lazzarato locates control as self-control and self-motivation. The cognitarian and semiocapitalist are the same self-valorizing and self-directed entrepreneur, although this overlap does not in any way exhaust the strange and troubling figures of semiocapitalists that haunt our world (on "terrorist entrepreneurs," see Genosko 2010). Cognitive entrepreneurialism was all the rage during the dotcom mania, and with the businesses created through the investment of intellectual assets like new search algorithms and plans for the next version of the World Wide Web, alliances with investment brokers and speculators were made. The luster of financial capital not only turned many beneficiaries of the new economy into market apologists (cyberlibertarians), but introduced the cheap pleasures of daytime and microtrading; of course, much of this crashed with the dotcom market bust between 2000 and 2002, and the acceleration of the real estate market, artificially buoyed by cynical subprime loans, which began to burst in 2007 (Marazzi 2010: 40).

Consider Berardi's (2008: 20–1) explication of Guattari's cryptic statement:

> "Capital is not an abstract category, but rather a semiotic operator." What does this mean? It means that while the labour process is fragmented, extended, recomposed and decomposed through deterritorializations of all kinds, the process of valorisation integrates all the fragments of capitalist production not only (not simply) through the abstract functioning of the laws of value, but also through the concrete, direct action of the technologies, allowing the instantaneous movement of information.

There are a number of points worth emphasizing. The first is that Guattari adds that such a semiotization is in the service of the capitalization of power, specific micro- and macro-social relations that assume any number of forms manifested relationally by individuals, groups, by rituals, objects, sexualities, ethnicities. These social relations of power, none of which are written in stone, but exploited by capital, refocus on the directness with which, Guattari insisted, semiotization relates to power and how this is manifested under local conditions.

Second, consider the fragmentation of the labor process or the becoming machinic of labor. Marx understood in the famous "Fragment on machines" of the *Grundrisse* (1972b: 693) that the automatic system of machinery, unlike the tool wielded skillfully by the individual worker, "is itself the virtuoso." Workers become accessories—conscious limbs—of machines. Labor is appropriated by and objectified in machines, to which living labor remains in a relation of subordination. This makes, Marx thought, machinery—which absorbs social intelligence and technical advances or social productivity—the "most adequate [objectively measurable] form of fixed capital" (1972b: 694). As the production process becomes progressively more technoscientific, and knowledge is objectified in machinery, the worker's activity is determined by the machines that have absorbed the "social brain." At the same time, capital might claim a false humanist dimension for the wage regime for, after all, workers are not themselves machines or fixed capital! (Deleuze and Guattari 1987: 457) Nonetheless, Marx's emphasis on the hard materialization of general intellect in fixed capital is being rethought in semiocapitalism.

Much recent work on machines has attempted to tease out the implications of the two processes described by Deleuze and Guattari (1987: 456–8): machinic enslavement and social subjection. In the former, workers are integrated and enslaved as component parts or pieces of a nonquantifiable machinic process and therein function as relays for fluxes, whereas social subjection still requires a certain distance in the subject's subjugation, as the user of a machine to which one is subjected, or category of person (salaried or subsidized). Lazzarato develops this distinction quite nicely by specifying that machinic enslavement works with affects, percepts, comportments, imaginations, desires, and a-signifying part-signs and places these flush with machines, while subjection is a more social or macro-scale identity, in which forced adaptation and alienation are evident and quantifiable. Lazzarato (2006a: 1) writes: "The machinic register of the semiotic production of Capital operates on the basis of a-signifying semiotics that tune in directly to the body (to its affects, its desires, its emotions and perceptions) by means of signs." Machinic enslavement and a-signifying semiotics work together. Lazzarato has what I would call a strong, contained sense of machinic enslavement as opposed to, yet consonant with, a more porous and weak sense of integration into networks; this may be contrasted with Massumi's (2002: 80ff) more porous use of the term, especially evident in his writing about transductive television—affective transmission and reconditioning of spaces of viewership.

Third, Guattari, too, notes a profound passage of human time into machinic time and automated info-technological exchange in which no measure can account for the blurring of the work/leisure distinction, nor for the "portion of machinic subjection entering human labour." Guattari observes that: "Automatized and computerized production no longer draws its consistency from a basic human factor, but from a machinic phylum which traverses, bypasses, disperses, miniaturizes, and co-opts all human activities" (1996b: 207). Berardi calls this the cellularization of labour: "Labour is the cellular activity where the network activates an endless recombination" (2009: 89). As production becomes semiotic, precariously employed cognitive workers—on occasional, contractual, temporary bases without guarantees or benefits—engage in labor that involves the elaboration of "a specific semiotic segment that must meet and match innumerable other semiotic fragments in order to compose the frame of a combinatory entity that is an info-commodity, Semiocapital" (2009: 89). Such segments or "semiotic artifacts" are highly abstract particles and they have a number of pertinent features:

- **i** they are fragments or partial, exiting before and below wholes;
- **ii** they are combined and recombined in a changing integrating framework ("the infinite recombination of a myriad of information" (2009: 88);
- **iii** their production and combination is intimately linked to the subjectivity of workers, thus making this kind of labor highly personalized, less and less interchangeable ("Consequently, high tech workers tend to consider labor as the most essential part of their lives, the most specific and personalized" (2009: 76);
- **iv** the production of part-signs and their recombination—connection, coordination and localization/planetarization—takes place on a machinic phylum: "The digital network is the sphere where the spatial and temporal globalization of labor is made possible" (2009: 89);
- **v** work-time is pulverized into atoms and reabsorbed by capitalist networks of production at the "exact moment" they are required; hence, "The cycle of production comes into operation only when it is required by the capitalist" (Lazzarato 1996: 136).

Simply put, an info-commodity under semiocapitalism consists in a nonexclusive way of a-signifying part-signs whose production and passage through digital networks contribute to the development of the machinic phylum which is, for Guattari (1996b: 126), the creative historical force of "selection, elimination and generation of machines by machines" engaging technological and living systems on a networked planet and describing them in both synchronic and diachronic ways. To inject Berardi's analysis here is to recognize that "capital no longer recruits people, but buys packets of time, separated from their ... bearers" (Berardi 2011: 90). The collusion between capital and the sign, cementing social roles and their required significations at the macro- or social level, and allocating unassigned affects and sensibilities at the micro- or machinic level, invokes the two processes of machinic enslavement and social subjection. Machinic components like part-signs help to develop the aforementioned phylum within their historical and logical entanglements, but Guattari emphasizes that there is "diagrammatic potentiality" at play in these knotted components of semiotic–immaterial formations. Not all of this potential is exhausted in being actualized but remains immanent, not as a reserve, but enfolded, "always looming," pace Massumi (2002: 226), "always, just." Surprises and autonomous shifts await construction according to new paradigms—like Guattari's ethico-aesthetic paradigm. Not everyone is so hopeful. As Berardi suggests, the general intellect is in a condition of ignorance in the space at the end of all futures, yet in this is a hopeful therapeutic poetry of failure that the cognitariat will reconnect with themselves: "We have to start from the ignorance of the general intellect. The force of collective intelligence is boundless. Theoretically. But it currently lacks any consciousness of itself. Intelligence without self-consciousness" (2011a: 163). "Fractalized work," Berardi (2011: 92) claims, cannot set in motion a wave of rebellion.

Under semiocapitalism, it appears that machinic enslavement has reached a peak of sorts given the integration of processes of subjectification and machinism in the info-webs of the machinic phylum. Berardi among others describes the "digital nervous system" and "brain-like network" in which cognitive labor is exploited: cell phones realize, for Berardi (2009: 90), "the dream of capital: that of absorbing every possible atom of time at the exact moment the productive cycle needs it. In this way workers offer their entire day to capital and are paid only for the moments when their time is made cellular. Info-producers can be seen as neuro-workers ... The entire lived day becomes subject to a semiotic activation which becomes directly productive only when necessary" (2009: 90). Here, he defines semiotic fragments temporally; this is the first of two key time-based comments he will make. The potential for escaping from the exploitation of semiocapitalism looms in the virtuality of the part-signs and autonomous affects relayed by workers who are enslaved by network life, propped up by pharmaceuticals, and primed for fielding commands to produce.

My central problematic is the relationship between semiotic and immaterial fluxes and how they connect under this semiotic machining of subjectivity. It is worth repeating, as Berardi tells us: "Semio-capital is capital-flux that coagulates in semiotic artefacts without materializing itself" (2007: 76). "Internal, mutual communication" (Deleuze and Guattari 1987: 458) characterizes the human-machine relays within expanding and perhaps generalizable machinic enslavement of the information age; put differently, semiocapital enslaves the so-called independent contract worker who writes code for, nominates names, or colors of, characters in video games (or perhaps writes and posts online cheats for favorite games); or, social media exploit for the purposes of data mining infrapersonal creative capacities, ultimately recombining many partial contributions into commercially valuable profiles, intelligence for advertising. What we learned from the discussion of a-signification was that semiotic–material "mutual engenderment" (Guattari 1996a: 151) was not subject either to substitution or cancellation by representational simulacra. The statements noted earlier by Virno, and Berardi, indicate that semiotic fluxes are not materialized or embodied, that is, do not appear to be anything more than immaterial. Yet something happens when semiotic products take shape. Coagulation without materialization or corporealization is the central emphasis of semiocapitalist production. This is what Lazzarato flagged as the inability to study, that is, to count, machinic enslavement which applies to production under semiocapitalism in which there are direct "recurrent and reversible" (Deleuze and Guattari 1987: 458) intimate relations between the semiotic capacities of cognitive workers and the framework of neuro-like webs through which signs circulate (some visibility is granted to those cafés where social entrepreneurs gather to stare at their respective screens throughout the day).

Varieties of immateriality

Critical insight into the coagulation of capital flux in semiotic artifacts expresses how perception itself can participate in the process of actualizing virtual part-objects. This was often discussed by Guattari in terms of theoretical objects, whose existence was only subsequently and retroactively confirmed or denied by experimental demonstration, but until then happily subsisted within a theoretical totality. Actualization does not require hard materialization; critical perception or conceptualization is sufficient to make non-material products slow down and pool, settling in semiotic artifacts. User affects, like postings of members of "communities of consumers, who co-produce innovation, diversification and identification with the brand" (Marazzi 2010: 58), help to create bonds among affects and intensities that cling to semiotic artifacts (i.e., the construction of "attention" or "interest" as an intangible asset in management theory). This occurs even among "hyper-material" commodities such as automobiles, some of which also have attractive and affective "cool quotients."

Both Guattari and Berardi emphasize that entire circuits and overlapping and communicating assemblages are engaged in the machinic integration of cognitive labor and the capitalistic exploitation of its content. What is this content? It is mental or non-material. When is it materialized? Mental as opposed to manual labor involves a closing of the gap between execution and innovation and a deferral of materialization; Berardi's (2009: 75) explanation contains an interesting qualification: "The materials to be transformed are simulated by digital sequences. Productive labor (labor producing value) consists in enacting simulations later transferred to actual matter by computerized machines." In this second, temporal qualification, labor thus loses its "residual materiality" in mental work on abstract "signs rich in knowledge" (2009: 76). This takes place in a networked environment, where communication (but not only this category because it is often separated from creation, and is as dull and repetitive as some manual work, that is, cyber sweatshops that specialize in building cheap web sites and designing info-products that serve as advertising vehicles) is worked with rather than being physically transformed; in the process, connectivity is enriched and extended. Flexibility and fluidity are imposed on such labor by the reticular formats (software programs, platforms, templates) that frame, capture, command, and recombine the fragments.

Several features of post-Fordist production are worth mentioning here. Industrial labor doesn't disappear, it is relocated to regions where wages are low and employment and environmental regulations are lax. Certain segments of cognitive activity follow along and are "externalized" for the same reasons. Accompanying Berardi's deferral, geographic marginalization of the moment of hard materialization is the automated support of the processes of recombination so that program languages, data formats, and the rest cohere and combine into a legible frame for an info-commodity that may need a material support. Two automatic systems have been alluded to, mobile devices (giving the necessary illusion of independence and self-enterprise) and techno-financial routines (illusorily autonomous but beholden to shareholder value) that impose rigorous fluidity and "determine continuous innovation" (Marazzi 2010: 65). These features engage a-signifying semiotics understood as part-signs of complex networks that trigger routines and subprograms and function by intimately connecting bodies to credit, spinning affects of trust, perceptions of access within personal banking paradigms, and contributing to the intangible sense of "confidence" that buoys investors, borrowers, and consumers.

Finally, as a theorist of semiocapitalism, Bifo forestalls materialization by means of coagulation in semiotic artifacts because he wants to underline the dependency of cognitive labor on information fluxes in the global networks of cellular telephony and the Internet. He does more than suspend the manufacturing moment; he is offering a variation, a detour, within the Guattarian relation between semiotically unformed functions and physically unformed matters. What he is suggesting is that "neuro-workers" (cognitarians) plug themselves into terminals of the network form, access a vast pharmacological support battery, and imbibe some of the fictions absorbed from the ideologues of the new economy. This precipitates widespread psychopathologies (Blackberry addiction), drug dependencies, and social dysfunctionality, not to mention personal indebtedness and reduced savings. Labor is pulverized into precarity, Berardi states (2012: 75). Berardi reflects on the "soul at work" by which he means not a representational substance but the "soul of Semiocapitalism," which is nonmaterial content, the very sort of semiocommodity under discussion here in the context of semio-digital ecologies, and how it may in some cases wend its way not so much back into but through other distributed nodes of informatized fixed capital. In this respect, he wants to pay close attention to how the machinic arrangements of fixed capital are evolving, and reacting to semioticization of the production process, in a mutating, artifactual ecology with long tendrils reaching across the globe; even so, partial manufacturing may be spread widely, and these semiotically unformed fragments coordinated by computerized assembly in another peripheral location: "the process of production is in large part dematerialized" (Berardi 2008: 20). To be fair, one may also say that production is in some part still materialized, which suggests that semio-material connective syntheses display a-signifying polyvocity and irregularity; while the disjunctive syntheses tend toward two extreme poles of a-signifying semiotization or remnants of industrial materialization, the conjunctive syntheses focus on the kinds and degrees of social subjection and subjectifications of labor.

What Berardi wants to find is evidence of the cognitariat's withdrawal from the semiocapitalist networks, and the shapes taken by its independent self-directed organizations since capital cannot really organize it: a cognitarian insurrection. The language of coagulation suggests, then, not a direct and immediate concreteness, but a soft, viscous, and centrifugal semiotics of actualization, displaying attention to the divergences of semiotic types, gradients of unformedness, variations between abstract and concrete poles, degrees of activity, container effects ("artifacts"). Coagulation of blood is a chemical process of clogging that requires the internal and active presence of platelets and specific proteins that become sticky. Most importantly, the theory of semiocapitalism can learn from the complexity of actualization, how it contains a measure of inactuality that makes it impossible to exhaustively grasp (Massumi 2002: 136), and suggests the fundamental fragmentariness of potentiality's manifestations and their mixing and remixing (recombination) through the capitalistic networks but in ways that confirm openness and multiplicity. And it is this always ready intervention that theorists of semiocapitalism want to regain from the capitalist exploitation of non-material products of immaterial labor, pulling out, redirecting and reapplying the potential "coalesced" countervalently adjacent to their semiotic actualizations, stealing it away from capital's semio-operations, from evaporating in the chimerical value production of financial capital and forging a social time distinct from capital time (Berardi 2006: 1). But coalescence is not coagulation as it is a process of merging that is abstract and temporary, in addition to requiring an external influence—like air currents merging water droplets, or consonants merging in a new phonetic articulation in our mouths. In other words, this is why Berardi likes WikiLeaks, Occupy, student protests, debt strikes, anti-austerity struggles, because all of these provide impetus for coalescence so that general intellect can find a social body that was denied to it by the conditions of semiocapital. Producers may coalesce but semio-products coagulate.

Of course, "coagulation" is a loaded term. Previously, I noted the use of it by Deleuze and Guattari as a species of consistency. In deploying the language of coagulation, Berardi recalls Marx's (1954: 47 and 64) explanation of value as the labor time congealed in commodities, but this is not an update. Rather, Marx's assumptions are shattered in the hypothesis of semiocapitalism:

COAGULATION/CONGELATION
Human labor (general social substance)
Labor time (average necessary)

On this view the "congelation of labor" as the substance of value that Marx analyzed is untenable because labor is itself more and more mental and digital, in becoming physically simplified it also becomes more personalized and hard to categorize; labor-time is atomized and indistinguishable from nonwork because anytime could be activated as work time depending on a project's needs, and this calls into question whether labor time is calculable in recognizable temporal units like a workday or hour. Indeed, rejections of the theory of value as relevant in describing social media point to the divergence of value and productive time as a measurable quantity: "labour ever more creates value in ways that are poorly related to quanta of time" (Arvidsson and Colleoni 2012: 140), but not completely since in some commodities labor time remains congealed as a "minor parameter."

Under real subsumption, time becomes a quandary as Negri recognized in *Time for Revolution*, as dialectics is surpassed and tautology and indifference result. Time is absorbed into the productive process and appears to measure only itself (Negri 2003: 27). Are average labor times knowable since every moment is under the command of capital? Are average skills still an issue when sensibility is the "skill" required to produce a semiotic artifact? Indeed, can value be measured at all? Is equal human labor coherent? Cognitive labor is deeply fragmented in time and space. Yet the "correspondence" (Hardt and Negri 2004: 68) between resistance and the dominant mode of production, cited in *Multitude*, yields:

Coalesce	Coagulate
Resistant (non) organization	Semiotic artifact
Temporary, mobile, singular, common	Soft and partial
Multitude, swarm, distributed network	Deferred or no materialization

The creativity of cognitive labor harbors a potential to be otherwise that coalesces as a tendency in the very process of coagulating semiocommodities. Coalescence coexists with coagulation. This paradox (Massumi 2002: 30; 226) is a feature of the virtual. If coagulation has a temporality of the not yet leaning toward an outcome, coalescence's temporality is closer to that of potentiality: "always just will have." Having arrived from no determinate place, that is, not from a quantifiable standing reserve, it comes as a surprise. As a nonresource, it cannot be exploited as such, but as a species of sudden arrival it resists specific renderings, yet holds much hope and difficulty to channel into, for instance, the multitude as a distributed network and the production of the common, or the refusal of work that cultivates new forms of autonomy, or molecular revolutionary perturbations at any number points of fractalized labor.

An archaeology of the semiotic coagulations of the immaterial that runs parallel to the potentiality of hopeful coalescences must seem for the reader of a Blochian disposition a return to a philosophical language couched in terms of another not yet (*noch nicht*) (Bloch 1970a, 1970b; Daniel and Moylan 1997; Hudson 1982). As resonant as the "not yet" appears across the Marxian and post-Marxian conceptual landscapes, one would do well to pursue with sobriety a twisting route between Bloch and Berardi since the nuances of expectation and their coloration in tones of process, experiment, and hope for the future, in the dynamic ontology of Bloch, while deep in Berardi's cycling around the poetic and therapeutic and his recourse to trying to understand happiness, is buried amid the bewilderment of depression, suicide, and the myriad subtractions that compose autonomy. Even if elements of the cognitariat like hacktivists and whistleblowers continue to find new ways to collectively outwit the institutions of security and foil the masters of secrecy by means of alternative connective coalescences, the struggle is far from over and hope tends to flicker, on and off the grid. Casting off the movement toward a higher reality to come (actual and true thing in itself) that guides Bloch, especially in his more religious pronouncements, and rather, embracing singularity, eventfulness, and the process of collective subjectification, Berardi, like Guattari, squarely faces chaos without a representational map and attempts to apprehend the tantalizing assemblies of virtuality that create the common, and then separate and fade like voices on the wind. "Not yet": another way of balancing between coalescence and coagulation.

Apprehending the multiple characteristics of congelation in Guattari's own work yields a simultaneous tension between pathology (passive) and confrontation with experimentation (active) in his construction of chaosmotic experience in psychotic patients:

CONGEALINGS
Passive: reductive, negative, subtractive
Active: emergent, affirmative, additive

Returning to the thematic of meaning's decentering and the role of a-signification that was discussed at length in Chapter 1 and revisited here, we can see one of the sources for Guattari's semiotics. As he explained in *Chaosmosis*: "we can see that a collapsus of sense will always be associated with the promotion of a-signifying links" (1995: 82) in which "signal-systems take off." Weaving together discursive and pathic components, Guattari's congelation is a trigger: for him, an implosive and depressive experience congeals and triggers considerations of how to rebuild a world for patients within an institutional space. He extracts from the congealed mass, from its non-sense, which only appears to be a vacuole but is enrichable, toward a manageable therapeutic outcome.

I mentioned Guattari's use of congelation earlier, but I would be remiss if I did not mention what Deleuze learned from the painter Francis Bacon: an aspect of painterly process involved the transformation of accumulated sensations from which the painter would extract forces, entailing that creation is not accomplished ex nihilo on a blank canvas, and is subject to unstable lateral flows from a variety of domains in the painter's studio, among his things and thoughts, etc. The painter turns matter into sensation, yet sensation is restless, vibrating; what is at issue is not, as Elizabeth Grosz explains, "the construction of pure and simple sensations, but the synthesis of other, prior sensations into new ones, the coagulation, recirculation, and transformation of other sensations summoned up from the plane of composition" (2008: 75). Deleuze (2003: 33) uses the term "coagulation," borrowed from Bacon, together with "accumulation" to characterize every sensation as a coagulation, "as in a limestone figure," from which the painter extracts and composes new composites by means of singular expressions (i.e., color combinations). Prior a-signifying traits or marks on the canvas are coagulated like so many givens with which the painter must reckon, even battle with, and transform. The synthetic unity of sensation in a given painting requires the painter to synthesize forces and fight his way out of the canvas. Thus, the painter must release himself from the gelatinous pool of accumulated sensations by means of random manual marks (wiping parts of the canvas, throwing paint at it) to clear away the clichés of representation among the givens in order to reach the intensive syntheses of a finished work. The manual marks are a diagram that fecundly melts the coagulations without causing a complete splash, and hence chaos on the canvas. Anticlotting is a necessity to ward against obvious figuration and to introduce by means of the diagram instability into the constraints of the identities of bodies on the canvas.

Fragments

Semiocapitalism did not crush Guattari's belief that a resingularizing combination of social experimentation and new technology was the best hope for exiting mass media and entering the universe of postmedia. The semiotic fragments produced by a precariously employed cognitariat may be detached from the dominant significations they are given by the abstract machines of immaterial production. Guattari (1995: 13) reminds his readers that these fragments "secrete new fields of reference" and in this rupture of sense "an existential singularisation correlative to the genesis of new coefficients of freedom will become possible." What curtails the capacity to apprehend and enter such new fields of reference is the devastating effect of semiocapitalism on the general intellect: semiocapital wants to subsume cognitive activity as such.

In tangling the Hjelmslevian net, Guattari prepared the ground for the investigation of how the abstract machines of semiocapitalism decoded subjectivity through info-machinic networks and the management of fluxes at the heart of immaterial labor and production, in the suspended space–time of the "not yet": recognizing the speeds and slownesses and species of soft substances, and the relative purities of form and im/material contractions. The plateaus of semiocapitalist subjectivation may be given names designating their singularities: Benetton, Berlusconi, etc. Guattari's own wily semiotic disobedience with regard to classic categories and theories helped to refigure how semiocapitalism attempts to enunciate the machine of human subjectivity by having it produce, enslaved and subjected to mobile networks, semiotic fragments in the manner of a-signifying particles in circulation. In other words, the direct exploitation of general intellect by semiocapitalism compels subjectivity to enunciate more and more like the nonhuman machines into which it is increasingly plugged and assembled but not, ultimately, into which cognitive activity is totally and infinitely subsumed. Yet the struggle against semiocapital's efforts to form subjectivity so that it hangs on a single designation of value and subsumes all the diverse types of valorization—aesthetic, desiring, ecosophical—to surplus value is, Guattari identified, the "deathly heart" (1995: 55) of all universes of value that draws them in and tries to obliterate them. As a semiotic operator, capitalism attempts to pump the implosion of value into the raw material of subjectivity, toxifying machinic inventiveness, transversal creativity, and devolving enunciative diversity. It accomplishes these by fragmenting time, boldly integrating into the machinic network general semiotic capacities, and making wages as minimal as the time mobilized transversally on demand.

The result of semiotic labor in the age of semiocapital is an info-commodity that need not achieve substantiation. Flirting with substance, the info-commodity retains a strange, gelatinous, de-solidity. Guattari's interpretation of Hjelmslev moved the analysis of production away from physical objects with determinate qualities, the socio-bio-physiological means of articulation, and social understandings of semiotic substances, toward productive creativity yielding an immaterial result (semiotic segments) that is combined with other such segments. The immaterial products of immaterial or mostly mental labor display both liquidity, which Berardi indexes to the liquefaction of time, and hence the semiotic flows keep the info-commodity from completely hardening, and temporariness linked to the unrelenting speeds of electronic information environments. Such environments facilitate the combination and recombination of abstract semiotic fragments by digital laborers who generate info-commodities by means of their highly personalized creativity—"mental states, feelings and imagination" across the surplus-value generating platforms to which they are "enslaved" as producers and consumers/users (Berardi 2009: 84). Semiotic segments can be put into place either by producers or users completing, in the words of Amazon's Mechanical Turk (https://www.mturk.com) marketplace for work on-demand, human intelligence tasks like copying, pasting, searching the web, and categorizing, or by cultivating successful social processes like sharing hyperlinks (i.e., among authors and audiences of microblogs). Lazzarato (1996: 144) reminds us that this is the "product" and it is "immediately the process of valorization"—of browsing habits, of linking to a site, of providing keywords for value and rank analysis.

4

Michel Foucault's Special Semiotic Characters: Obstacle-signs

In Foucault's *Discipline and Punish* (1977), there is a prepanoptic discussion of an art of signification that places representation at the heart of the public communication of a highly diverse array of punishments. Within the logic of exposition, Foucault's discussion of the "gentle way of punishing" precedes the production of "docile bodies" and the improbable emergence of imprisonment in a "coercive institution" as a general form of punishment; in the process, the "picturesque"—to use Rudy Visker's (1995) emphasis—penalties previously imagined by reformers are superceded. Foucault's question is, how did prison detention become "one of the most general forms of punishment?" (1977: 120) In this chapter, I want to reframe this question: what was lost when the prison became "one of the most general forms of punishment?" What, in short, can be learned today from a semiotics of reform in which representation and not coercion of the individual through a "concerted orthopaedy" is the goal of the penalty; where signs are not displaced by exercises, and the locus of the power to punish includes special semiotic characters that perfuse a so-called punitive city? Regaining this semiogram of power will contribute to bringing a Foucauldian meditation to bear upon the analysis of the neoliberal surveillance state; still, this is perhaps too broad for there is in Foucault a city of sorts at stake. My specific goal is to extract and describe the features of special semiotic characters called "obstacle-signs" (*signes-obstacles*) from Foucault's account of what he calls a "dream" of penal reform dating from the eighteenth century that died and gave way to the "great uniform machinery of the prisons" (1977: 116). The diverse signifieds of obstacle-signs were reduced, that is, "colonized," by imprisonment, following the disappearance of public torture and the dissipation of the whiff of abuse and illegality that prison previously carried.

My goal is to show the enduring relevance of obstacle-signs as a semiotic invention for critically apprehending our current situation, at least since the 1980s, within a burgeoning neoliberal condition, but whose example will be focused on the link between drinking and driving. But first, a certain amount of explication of Foucault's text is in order, to be followed by a reflection on the semiotic character of the obstacle-sign and the contribution it makes to a critical semiotic theory by departing from the principle of arbitrariness.

Semio-techniques

Foucault describes how changes during the second half of the eighteenth century regarding property laws and the development of capitalism in the forms of intensive agriculture, investment in manufacturing, transportation, and port expansion provided the bourgeoisie with privileged domains suitable for working through its political machinations. Notably, these reforms also effectively introduced a "new economy" of punishment based not on spectacle, confused principles, and haphazard applications but on "continuity and permanence" (1977: 87). This "new economy" required individuals to accept a contract with society; those who break this contract through criminal acts become enemies of society. The part and whole are related in this way: the least crime attacks the whole of society, and the whole of society is present in the least punishment. In this way, punishment is generalized, Foucault underlines. This struggle between part and whole is "unequal" since the power to punish belongs only to society; vengeance should play no role. Instead, the question of punishment is the defense of society and the expression of the humane sensibility of the lawmakers, yet implying a principle of calculation deep within it that attends to the recoil effects of punishment upon those who exercise it; attention is given, in other words, to the "effects of power" (1977: 92). Punishment is an "art of effects" (1977: 93). It is a matter of economy and proportion formulated to "prevent repetition" by perpetrators and their imitators. Reference back to the crime in the punishment must be "discreet." Punishment is not an example that allows for the reappearance of either crime or sovereign power; punishment, therefore, is a punitive "sign that serves as an obstacle":

> Through this technique of punitive signs, which tends to reverse the whole temporal field of penal action, the reformers thought they were giving to the power to punish an economic, effective instrument that could be made general throughout the social body, capable of coding all its behaviour and consequently of reducing the whole diffuse domain of illegalities. (1977: 94)

There are six major rules of this semio-technique:

(1) Minimum quantity: Reduce the advantages that a crime creates for the criminal by stripping away what is desirable about it while creating greater interest in avoiding the penalty;

(2) Sufficient ideality: Given rule 1, the "pain" of punishment must function as an ideality, an "idea" and not a "corporal reality," a representation, which punishment maximizes;

(3) Lateral effects concern the centrifugalization of the effects of punishment: The effects of punishment spread centrifugally and must be managed. The goal is to "make others [other than the criminal] believe that s/he [the criminal] has been punished" (1977: 95), and to do so in a lasting way;

(4) Perfect certainty: Constitute by means of representational signs a "necessary and unbreakable" (1977: 95) link between the crime and its particular punishment. This entails a clarity of consequences delivered through the publication of written legislation and the speeches of magistrates; open access to records should also be guaranteed. Here Foucault introduces an important feature: in order to realize the perception of certainty, "no crime committed must escape the gaze of those whose task it is to dispense justice" (1977: 96);

(5) Common truth: Verification of the crime must obey accepted principles of truth in the arguments, proofs, and judgments that support it. This requires a transition from "ritual" acts to "common" instruments ("reason possessed by everyone") in which one is innocent until proven guilty (1977: 97);

(6) Optimal specification: In "penal semiotics" (1977: 98) illegalities must be rigorously and comprehensively defined and coded, classified, categorized in a taxonomy of punishments, with no excesses or loopholes: the goal was an airtight semiotics in which there would be a "total coincidence between all possible offences and the effects-signs of punishment" (1977: 98). Additionally, the "criminal himself" must be considered so that punishment may be "individualized" in a "dyad" of mutual reinforcement (this was still, during this period, "far off" in the future); yet, "individualization appears as the ultimate aim of a precisely adapted code" (1977: 99).

Punishment could be "modulated" in a new way that was different from the prior modulatory casuistry according to the defendant's character, way of life, history, quality of will. The code and individualization scheme were developed in relation to natural history taxonomies, specifically the Linnaean taxonomy. The code was portrayed in a great table, a spreadsheet, divided into crimes according to their objects and then into gradations of species, from which comparisons to another table of penalties could be made. This "double taxonomy" was more of a "dream" than a reality (1977: 100), a fantasy of exhaustive analysis.

"Modulation" is a rich philosophical term and its use by Deleuze in his late writings on the control society is worth commenting on. Foucault introduced the term in the mid-1970s and periodized it in a way between its application to individuals in terms of doctrine as opposed to a later idea of the psychological impact of punishment on individuals. Indeed, Foucault excavates an unrealized dream at the end of the eighteenth century in the form of a penal semiosis that could be modulated so that it touched the "minds of all"— not the bodies of one and all, but the souls of everyone by means of representative signs, the so-called obstacle-signs. Such a new semiotic gentleness would eventually give way to the application of punishments in a "new politics of the body" (1977: 103). It is widely believed that Deleuze went beyond Foucault, and even "corrected" him (Elmer 2004: 43) by displacing disciplinary societies based on confinement with control societies that capitalize on the breakdown of institutional sites and the introduction of free-floating, open systems based on modulation (alterations according to circumstances) as opposed to molds (firmly set and into which one is made to fit). Deleuze (1992) aligned modulations with change, self-transmutating and self-compensating, linkages, sharing information, short-term time frames, flexibility, mobility. It is evident that modulation was already at play in Foucault's *Discipline and Punish* in his reflection on semio-techniques, and that Deleuze breathed new life into the concept. With Foucault, it may be claimed that his analysis of the prospects of the effects of the "play of representations and signs" directed at the entire populace introduced diversity and flexibility into the nightmare of an urban semioscape of punishment, not unlike the post-Orwell surveillance state of "total information awareness" or "ubiquitous surveillance," and the current goal of the National Security Agency in the United States to capture *all* communications. The proliferation of representations that Foucault described, which I discuss below, does not quite advance toward the destratification of exclusionary institutions and disarticulation of hierarchies (Bogard 2006: 59), but it does suggest diverse immaterial semiotic forces (signs, ideas, performances, publicity, poems, "picturesque punishments"). The difference between disciplinary and control societies, Bogard (2006: 63) writes, is like ice melting into water: "Discipline becomes liquid: it flows into every hole, fills every crack, and leaves nowhere to hide."

Obstacle-signs

Transitioning, then, from generalized punishment to the "gentle way in punishment," Foucault explores the role of the "technology of representation" in punishment through what he calls "obstacle-signs" (1977: 104). The role of representation is to rob the crime of its attraction by communicating its disadvantages by means of such signs, thus constituting a "new arsenal of penalties" (1977: 104). The six conditions under which these types of signs function are as follows:

(1) Avoid arbitrariness as much as possible; be "as unarbitrary as possible" (1977: 104). Such signs are "analogical" (based on resemblance and proximity) in design in as much as a specific punishment is immediately and transparently signified by the thought of a crime, the signified acting as a deterrent that "diverts the mind" from going down the road of that commission: "the transparency of the sign to that which it signifies" (1977: 106) is aimed at all those experiencing the representation. The semiotic relation is between signifier (crime) and signified (punishment). The link would be immediate, stable, transparent, natural (and as such "gentle") as far as its "sequencing" and "force" are concerned. This notion of the natural sign meant that vainglory would be punished by humiliation, murder by death, theft by confiscation (1977: 105). In other words, punishment underlines the "symbolic communication" between the crime and its "analogical penalties" (1977: 105). Today, in some constituencies, chemical castration for repeat sex offenders and the death penalty for serial murderers appear to retain this symbolic principle of communication. Below, I will emphasize the sobering transparency of the relationship.

(2) Signs engage with and subdue, weaken, dissipate forces of desire, attraction, and sundry intensities and pleasures of crime. Force is marshalled against force, good habits against bad passions: it is advised to "set the force that drove the criminal to the crime against itself" (1977: 106). In this sense punishment is an irritant that upsets the pleasures and rewards of criminality. But this is not only a negative operation. Additionally, Foucault describes how what is useful and respectful must be rebuilt, "reanimated" so that "the penalty that forms stable and easily legible signs must also recompose the economy of interests and the dynamics of passions" (1977: 107). To the transparency of obstacle-signs is added stability and legibility: ease of decoding.

(3) The idea of the penalty involves a "temporal modulation" (1977: 107), that is, it is not permanent. Torture was no stranger to temporal considerations, and duration became a way of transformation rather than an ordeal. Diminishing intensities and greater leniencies are thus imagined as internally variable yet nonetheless fixed sequences. Again, Foucault's recourse to the concept of modulation is in the service of describing the penalty's variable durations and diminishing phases and how the subjectivity of the convict would be modified.

(4) Punishment is not only directed at the convict: for him, the penalty is an obstacle-sign with a tight fit between crime and punishment, but the reach of the sign extends beyond the guilty to the "potentially guilty" (1977: 108) and is meant to circulate widely and rapidly, find acceptance and redistribution, shaping inter-individual discourse about crime's falsehoods and its inevitabilities. In this way public good can be made of convicts. They provide physical labor and perform a semiotic service "by the signs that [they] produce" (1977: 109). Semiotic action is required against crimes yet to be committed and crimes not yet conceived of, but which would be decodable within this penal semiotics and the coordinates of the taxonomical spreadsheet.

(5) Terror may have been engraved on the memories of those who witnessed public executions, but obstacle-signs are "decipherable" by all. This results from the "learned economy of publicity" in which obstacle-signs deliver unambiguously decodable moral lessons. They radiate a clarity of message evident to all. Another "double" emerges: the double affliction of the citizen who ignores the law and becomes in the process a criminal "lost" to society. Sadness and mourning rather than vengeance are the emotions proper to this ritual; the effect is not terror, but a legible publicity of punishment: "it must open up a book to be read" (1977: 111). Here is a great social semiotext: the flying of flags depicting "penal heraldry" with careful color-codings and bestiaries reflecting crimes and their punishments. This book, "ever-open," belongs to the ubiquitous school and not the festival initiating a kind of continuing education.

(6) In making criminals into sources of instruction, the "dubious glory" of criminals may be extinguished. This requires constant reinforcement by the circulation of obstacle-signs. The "positive mechanics" of which recode punishment so that the pleasures of crime, the nobility of outrageous exploits, the mastery and honed skills of the finest thieves, are diminished and crowded out in popular discourse by fear of punishment: "Instead of those songs of praise that turn the criminal into a hero, only those obstacle-signs that arrest the desire to commit the crime by the calculated fear of punishment will circulate in men's discourse" (1977: 112). Discourse as the vehicle of punishment captures poetry, popular tales, folklore, the moral lessons that parents tell their children: the imaginary "punitive city" is perfused with public obstacle-signs. Everywhere is a "tiny theater of punishment" that plays and replays, through elaborate devices and effects, the semiotics of punishments (highly visual), in all-ages shows of nonstop moral education: "The great terrifying ritual of the public execution gives way, day after day, street after street, to this serious theater, with its multifarious and persuasive scenes" (1977: 13).

According to the logic just presented, imprisonment or detention would correspond to a crime such as kidnapping or hostage taking. It is the sort of punishment that suits a specific type of crime and is analogically related to it. It is semiotically coherent: "it will be important to avoid utilizing the same punishment for different crimes—which would be like using the same word to say different things. Instead of putting lazy vagabonds in prison, put them to work" (Ransom 1997: 32). In other words, imprisonment lacks specificity when used generally and it is criticized in this way by reformers who believed it to be counterproductive: it turns semiotic abundance into a single disciplinary enclosure. Foucault writes, "prison as the universal penalty is incompatible with this whole technique of penalty-effect, penalty representation, penalty-general function, penalty-sign and discourse" (1977: 114–15). Still, by 1810 (Foucault largely derived his remarks from texts written in the 1780s and 1790s), prison occupied "almost the entire field of possible punishments" (1977: 115). In all of its manifestations, prison was a "monotonous figure… of the power to punish" (1977: 116), a bloated signified that severely restricted the representational field of publicizing punishment. The widespread "practice of imprisonment" itself comes to be seen as natural despite the astonishment of those reading about the aftermath of the efforts of reformers during the late eighteenth century (Foucault 2002: 225–6).

Special characters

Obstacle-signs are framed in the terms of a conceptual language with which we are familiar from structural linguistics, but the specification of the conditions of signification are divergent: the relationship between the signifier and signified of an obstacle-sign diminishes in arbitrariness as resemblance, analogy, and proximity are foregrounded as the principles of motivation governing the system of crimes and punishments. Obstacle-signs need to be rendered starkly transparent and intelligible so that the unity of the signifier and signified cannot be shaken or diverted by the affective charms of crime or the hope of nondetection either through a blind spot of signification or a fuzzy polysemy. This requires reinforcement through a comprehensive and intensive mediatization—the "ever-open book" to be read by all. Thus, such a semiotic must be sobering in its clarity and never casual in its ineluctability, yet be delivered through a landscape perfused with "tiny theaters of punishment" where written discourse and oral performance deliver the strictest of lessons intergenerationally and semiotic disobedience is minimized. Obstacle-signs must arrest desire. Therein representation is not at all festive and engages multiple media as well as linear textual and visual modalities (posters and heraldry). Recalling Jean Baudrillard's (1993a) discussion of the sure signs of a symbolic order prior to the emergence of modern signs marked by counterfeiting and arbitrariness, obstacle-signs answer to this transparency of certainty and circumscribed circulation as lessons in prohibition, acceptance, and submission. Such obliged signs are nonarbitrary or motivated. In Foucault's description, however, obligation and surety are goals of the theaters of punishment rather than guaranteed or given by the kind of society imagined.

Obstacle-signs are positive in the sense that they are defined by the fullness of the motivated relations between signifier and signified rather than solely by the negative differences between signifiers and their meaning effects as the source of their meaning within the penal code. The translation of the recurring phrase *jeu de signes-obstacles* by "complex of obstacle-signs" (Foucault 1975: 106 and 108; and 1977: 104) allows the weight of an ensemble as hybrid coding to come through but without resorting to the more structural language of system articulated by means of the theory of value. The signified is not an effect of the structural interdependence of signifiers. Each obstacle-sign is ideally quite specific in its construction; the uniformization of the penalty by prison effectively destroyed the internal diversity of obstacle-signs. Recalling Fredric Jameson's (1991) analysis of postmodernism in which the interdependency of signifiers "snaps," revealing the "rubble of distinct and unrelated signifiers," Foucault's description of a premodern semiotic imaginary is likewise snapped over the course of some twenty years (approximately between 1790 and 1810) in which the signified is banalized and monotonized as a single kind of penalty that is no longer "telling" in relation to the specific offense committed. While Jameson describes the schizophrenic consequences of the snapping of the signifying chain in which a subject is "engulfed" in the "vividness" of an isolated signifier heavy with affect, we may see in the transformation of the signified as imprisonment, the general form of punishment, not only an evacuation of variation, but a stripping away of the communicative effect in which a subject no longer confronts obstacle-signs that warn against specific criminal activities at every turn. Rather, with the disappearance of the public signscape of crime–punishment scenarios, and their multiple activations of fear and acceptance, this movement is also a desocialization that recedes into the coercive institution of prison. Public signs are replaced by tightly controlled architectural enclosures.

The generalization of the signified of imprisonment also introduced redundancy and imbalance into the vision of a Linnaean taxonomy of crimes and punishments. Even such a reductionistic social classification scheme, emphasizing as Foucault (2002: 363) observed, techniques of spatialization and visibility, would be hard-pressed to accommodate a comparative table for penalties dominated by one outcome such that any complexity of correspondence between the gradations of crimes and punishments tables would be radically curtailed. The signifiers (crimes) would remain vivid yet lack unique analogical signifieds (punishments). Instead, the signifieds would become passive, ordinary, and common—less and less "picturesque" and considerably duller. A further feature of an imagined taxonomic table regains Foucault's (1972: 67) sense, in *The Archaeology of Knowledge*, that discursive constellations of this kind are "essentially incomplete" and his example is Linnaeus's "free" elements, which by extension suggests that there remains room for new, intrinsic, punishment-types to emerge in expanded or supplementary tables.

Beyond the generalization of the signified, and the ease with which imprisonment may be found, is the inability to find a signified. A loosening of the analogical tightness of the sign may be described by the intrusion of an obtuse factor, what Barthes (1977: 54) theorized as an excessive and elusive supplementary dimension of meaning. Such indifference to obviousness challenges the clarity of the obstacle-sign, and the extreme result may be a "signifier without a signified." Thus, in Foucault's terms, a representation that fails to communicate a specific punishment has broken down. It is not necessarily bad theater, but a theater not in control of the effects of its representations and unable to manage the accents emerging from it. Foucault does not acknowledge that among the hundreds of tiny theaters attempting to communicate to the "potential guilty" that there must have been, in this vision of reform, not only aberrant significations but challenges to the machinery of representation. For the obtuse, as Barthes explains, resists description and representation. Generalization of the signified and obtuseness belong on the same continuum. At the other end of the continuum is found the breakdown of the analogical relation and the emergence of purely arbitrary relations. With reference to Saussure's insistence on the arbitrariness of the sign, there is "no specific agency" (Bouissac 2010: 96) responsible for lining up acoustic images and concepts; however, in Foucault's descriptions, the inculcation of analogical signifying relations in the populace presupposes agencies of some sort and if these showed signs of failure by having widespread recourse to one dominant signified, blame could be apportioned and their representations could be dismantled. This situation is precisely the opposite of the one faced by Saussure in distinguishing motivated from nonmotivated signs, situating arbitrariness outside the field of assignable, agential influences on signifier–signified relations. Indeed, it might appear to be beyond Saussurean considerations of motivatedness as such, even in the "relative" forms of arbitrariness that he introduced (Saussure 2011: 131). Perhaps the waning of analogical signification is nothing more than an example of the "see-saw" movement that Saussure identified between arbitrariness and motivation, and we can reverse Saussure's dictum that it is impossible to have a purely arbitrary language without any motivation (see Harris 1988: 54), so it would be impossible to have a purely natural or motivated one as well without some arbitrariness. But this is not really necessary because after all we are not dealing with only language in Foucault's obstacle-signs. While a number of scholars have criticized Saussure for his introduction of flexibility around arbitrariness, considering it to in one manner or another diminish the systematicity of his theory, there is little to worry about, for others, since it is not possible to predict the derivational patterns from among the many alternatives that are possible for signifiers and signifieds presenting some rationale for their bonds (Gasparov 2013: 75). In the semiotics Foucault is rendering, outbreaks of arbitrariness are eruptions of desire.

"Hundreds of tiny theaters of punishment"

Within the neoliberal diagram of power, governing through wars on crime is a standard conservative strategy. The dream of a public and "permanent lexicon of crime and punishment" (1977: 111) suggests former US president Ronald Reagan's utilization of anti-drunk driving legislation in the early 1980s (see Lerner 2011). During this era, federal legislation normalized blood alcohol content levels, enshrined license suspension and mandatory penalties, and established better law enforcement. The National Minimum Drinking Age Act was also passed with the support of groups like Mothers Against Drunk Driving (MADD), which advocated for victims' rights and found a fertile political and media environment for its cause. MADD established chapters throughout the provinces in Canada as well. If we consider the desired effects of obstacle-signs to reach beyond the convicted to the "potentially guilty," then the ways in which they shape discourse, annotate, and even revise the book to be read, as it were, is especially pertinent to understanding how these sign-types that Foucault surmised from his analysis of eighteenth-century texts might be relevant for the writing of a history of the present.

In the case of drunk driving, the link between the presence of a blood alcohol content or concentration at or above the minimum (in the province of Ontario, it is 0.08, 80 milligrams in 100 milliliters of blood) is the single causal factor explaining from 25 to 50 percent of car crash fatalities; this cemented the signifier–signified relation in which the crime of drunk driving (blood alcohol level) and its punishment (analogically understood as a suspension of one's license to drive, and later, in terms of proximity, cars that test such levels and block ignition) circulated in mass media, and by government, and exercised a remarkable moral authority, for a short time. The publication of the names of those convicted of driving while intoxicated reveals what Foucault considered a double payment of the convicted: the consequences of a DWI (driving while intoxicated) conviction and the production of obstacle-signs for those who might consider driving (or operating another kind of vehicle like a boat) after or while consuming any amount of alcohol; ultimately, this would bleed together impairment and driving under the influence. Of course, the overdetermination of the idea of "drunk driving" (as opposed to driving in the so-called warning range with BAC between 0.05 and 0.08) and the isolation of the single most important cause of car crash fatalities as blood alcohol content (as opposed to alcohol's behavioral effects on driving ability) were criticized by multifactorial causal analysis and re-examinations of the myth of the drunk driver as the "villain of traffic safety" (Lerner 2011: 97). But this villain, once put into circulation, effectively "recoded" drunk driving as the act of an "enemy" of society, assisting in the diminishment of any value accruing around the desire to engage in such activity, and therein limiting appeals to alibis such as tolerance for social drinking and recourse to populist nostrums like having "one for the road" as a poor person's tonic for the nerves. By putting the "killer drunk" at the center of its "moralistic analysis and demonology" (Heron 2003: 343), the women of MADD staged populist theaters of punishment whose performances proliferated in the forms of police check stops, designated drivers programs in public bars and private homes, roadside license suspensions, advances in "breathalyzer" and ignition lock technologies, public health, and safety education. Indeed, sentences handed out to drunk drivers are regularly publically debated in the media by victims' families who often question the tightness or looseness of the fit between crime and punishment.

Despite this semiotics of sobriety, a link between drinking, pleasure, and sociality persists in the popular imagination because alcohol still forms part of a preparation kit for driving under certain, perceived-to-be-difficult circumstances (e.g., driving at night, in the snow and ice). In fact, if alcohol forms part of a healthy lifestyle, which is occasionally demonstrated, it must be read together with its dangers, resulting in ambivalence that weakens the crime–punishment bond (Heron 2003: 386).

The anti-drunk driving "serious theater" of the punitive city is perhaps best witnessed in the MADD red ribbon and other campaign materials, liquor control board moral messaging (e.g., by the state-operated liquor control boards in some of the provinces in Canada), and holiday clampdowns at the roadside by police. These campaigns attempt to lock-in the signifier–signified relation of impaired driving-license suspension and/or monetary penalty by scare tactics or compelling theater. The signified of driver license suspension is a "telling" punishment, even if it cannot be described as particularly picturesque. However, the installation of an ignition locking device—so-called alcholocks—for repeat offenders is a more dramatic kind of performance; so, too, is a roadside license suspension and car towing. Since the 1980s one may see in the penalty for street racing—impounding vehicles—a similar obstacle-sign of a "safe roads" code working in concert with other driving penalties. What I am maintaining is that for a brief period in the 1980s the legible lesson of the crime and punishment ligature was tightly wound in the spirit of obstacle-signs, even though over time the problem of the common, dull signified of license suspension would only be partially addressed in terms of innovations that emanated from the vehicles themselves with built-in breathalyzers, other alcohol detection systems, and ignition locks. The focus of the 1980s gave way to confusion among different constituencies (states and provinces in the United States and Canada) about the language of driving while intoxicated and driving under the influence (DUI is known in Canada as impaired driving, yet it persists in the common language as it is undoubtedly inspired by television cop shows).

Obstacle-signs appear to be "naturally" politically conservative and support a moral mission through multimedia representation, what we now call in Canada a "punishment agenda" and "governing through crime" in the United States; conversely, the political use of public drinking by a subculture to carve out a space for itself is arguably restricted to similar conservative outcomes, in addition to the proliferation of obstacle-signs (Marcus 2005: 271). "Law and order" campaigns tend to give up their diversity for the generalization of punishment in the form of prison and mass incarceration. The eighteenth- and twentieth-century examples share this generalization of punishment. Because they communicate the kind of "certainty" that reduces what Foucault terms the "coefficient of improbability" (1977: 96) that otherwise loosens the bond between crime and penalty (the ability to avoid roadside checkpoints), a further apparatus is required. Surveillance—the "gaze of those whose task it is to dispense justice" (1977: 96)—is the means by which this reduction would take place in support of the surety of laws, their publication, and advertisement, yet Foucault only pursues the issue of the gaze (single-normalizing-permanent) within the context of disciplinary institutions, that is, after dispensing with obstacle-signs. It is less remarkable that the gaze appears within the discussion of obstacle-signs than the loss of the explicit study of signs (not the power of ideas and diagrams) in Foucault's subsequent treatments of the contributions of the human sciences to the prison's productivity in its fabrication of individuals (delinquents) as effects of power/knowledge. The contribution of signs to epistemological power appears limited to a brief and somewhat dreamy historical moment, and quickly pales beside psychiatry and criminology, and is thrown out by Foucault with the bathwater of "structure."

Foucault does not concern himself with the semiotic vicissitudes of the punitive city. Further, even though he has at his disposal an understanding of the unwieldy nature of public corporal punishment based on his analysis of the theatricality of public torture and the political problems created by the public's behavior (making the criminal a hero, mocking the sovereign, generating new and dangerous solidarities among an undifferentiated mob), he does not transfer and apply this to the "punitive city." This is undoubtedly due to his inclusion of semio-techniques among all of the refinements of penal reform; that is, all of the ways to make punishment more subtle, more logical, more balanced, more qualitatively nuanced. Hence, a shift from the referent to the sign, from the real to the ideal, in the (re)presentation of crime and punishment: the theater replaces the scaffold, semiosis substitutes for ritual.

As I have suggested in this chapter, obstacle-signs are special semiotic characters that have a more recent history, remain in circulation today, and are well adapted to specific political climates and temperaments for the development and deployment of reactionary regulatory regimes. The survival of certain obstacle-signs does not mean the "punitive city" of dispersal and diversity of control has been again realized, even in the most theatricalized "gradations" of community correction/treatment schemes (Cohen 1979). These surviving signs carry with them, however, obstacles to a future made tense by the perceived need to prevent its inevitability, both by the potentially innocent (exonerated guilty) and by their nefarious others. The red ribbon campaign that MADD created festooned the mediascape of the 1980s with a strong bond between the signifier (red ribbon) and signified (commitment to sobriety). The red ribbon was an arbitrary sign for neo-temperance values. However, the MADD goal of eliminating drunk driving across North America generated numerous obstacle-signs that the organization performed with the help of a very deep budget built upon successful fundraising activities, media-savvy representations, and a political climate receptive to law-and-order issues and victims' rights agendas. The success of the obstacle-signs in circulation may be seen not only in reducing drinking and driving for fear of the inevitable punishment that awaits the driver after blowing into the breathalyzer in a cityscape perfused with surveillants, but that the message reached those who served alcohol, in public establishments and in private homes, and that new actors like the designated driver took a leading part in the dramas. It is perhaps fitting to point out that the book to be read by all is now more comfortably a floating yet well-established awareness of a natural bond between drinking and driving and some predictable, highly inconvenient, and socially stigmatizing, not to mention costly, form of punishment.

5

Jean-François Lyotard's Tensor Signs and the Passage to Affect

In this chapter, I will revisit Jean-François Lyotard's provocative introduction of an energetics into semiosis in the guise of tensor signs. In his challenging book *Libidinal Economy* (1993), these signs foreground unaccounted for and undertheorized movements, inflations, and deflations, at times vertiginous and unstable relations, in a tradition of sign studies often at odds with affect. This amounts to a critique of rationalist semiotic conceptual conservatism and denial of sensuousness. Libidinal economics is Lyotard's unraveling of discourse and structure; likewise for Guattari, the machine was a disruptive object introducing disequilibrium into structure and was modeled on psychoanalytic part objects. It is difficult to follow to the letter Bill Readings in his *Introducing Lyotard* in labeling Lyotard a deconstructive thinker. Readings is aware of the pitfalls of this claim, and despite citing some of Lyotard's objections to Derrida in several books, sticks with the label, despite himself. I offer a different approach, with occasional comparisons with Derrida, based on how Lyotard's heightening of intensity opens the way to the rise of affect and the challenge it presents to semiotics in general, to which I turn in the second section of this chapter.

Over the course of two theoretical works, *Discourse, Figure* and *Libidinal Economy*, both originally published in the early to mid-seventies, Lyotard's challenge was in part a critique of signs and representation, in a manner similar to Guattari; as Lyotard put it in the former book, "the ambition is to signify the other of signification" (2011a: 13), what he called the disruptive figure that is accessed in discourse; whereas in the latter book, he sought to think intensity as structure, not to oppose intense to other signs, but to think them together at the same time. Both the figural and the tensor, just as the machine, are disruptive of language understood semio-structurally. There are questions that must be faced for Lyotard's readers, as his appeals to pure intensity not only raises the specter of fascism (a scarecrow, says Lyotard, in dismissing it), but with regard to his own construction of semiotics, while not reducible to specific straw men, is selective in its examples, invoking both Derrida and Baudrillard toward the construction of a semiotic nihilism and its "game of de-presence" (Lyotard 1993: 46). Throughout this chapter, I will remark on some points of convergence between Derrida, Deleuze, Guattari, and Lyotard on the question of force in semiosis. The key foil will be Saussure and his semiolinguistics.

Tensor signs

The great Moebian (folded yet one-sided) libidinal skin or band that Lyotard theoretically unfolds at the outset of *Libidinal Economy* is a field of intensities, sometimes called forces, subject to decay by means of devices that carve out interiorities (representational theatrical chambers) and therefore brace themselves against some outside, and in so doing threaten the skin's pure affirmativeness: "The operator of disintensification is exclusion ... the disjunctive bar" (1993: 14). Inside or outside, this or that, but not both at the same time. Not-this is the exterior negative relationality of the Saussurean sign that cannot be treated in isolation. Lyotard's penchant for "this or that" type disjunctions raises the question, from a Derridean perspective, of whether there are minimal and maximal states of absence; Derrida suggests that pure absence differs from this or that (1978: 8). The concepts of difference and arbitrariness are two ways of slowing down the spinning bar. As Readings (1991: 10) once joked, the sign in Saussure is a hole because what it is is really as much what it is not, maybe even more so if any consideration of the sign's body is discounted. This is an apt Lyotardian joke as for him the slowing down and cooling of the aleatoric movements of the spinning disjunctive bar that assists in producing the aforementioned skin (the latter doesn't simply preexist the bar, but chicken and egg questions bored Lyotard) leaves in its wake set-ups marked notably by their emptiness (in the guise of lack, things lost, nihilistic decompressions, boundaries that need to be negotiated, and the discriminations to which they give rise). Perhaps Readings's "hole" has a finer coating put on it by Patrice Maniglier's (2010) insightful discussion of how semio-structural displacement, the "logic of the signifier," was utilized by Jacques Lacan and members of his psychoanalytic school to maintain that the positionality of the sign, which included its own absence, could as a principle allow the subject's non-self-identity to support (guarantee) every term's non-self-identity in the signifying chain. Further, the subject as an effect of relationality and the lack of properties (Saussure's "without positive terms") of terms in the negative differential structure was a move that vexed Deleuze in *Difference and Repetition* in his thinking of plurality and positive multiplicity against the reduction of difference to opposition, especially by linguists (1994: 204). As an antistructural masterpiece, one could claim that the very thrust of Deleuze and Guattari's *Anti-Oedipus* (1977: 309) is to find a "pure positive multiplicity": not negative, not lacking, not answerable to difference; but full of possibilities, transversal connections, and a-signifying terms in the chain, all squirreled away in the deepest recesses of structuralism, no less. It is worth noting that Lyotard, in *Discourse, Figure* (2011a: 25), recognized that the Saussurean sign and law of value contain a specific force of negativity (negative difference), but it is not the only "no" he will work with; for instance, structural linguistics' deferral of designation (2011a: 50) and the inability to account for reference are regained as instances of the introduction of figural visibility into the otherwise circumscribed system. I will return to this strategy later in the discussion.

It is helpful to ask Derrida about the conditions affecting force and signification: for him, force is defeated by structure, or at least neutralized by it, even if this too is a force of sorts (1978: 4). In Lyotard's philosophy, a sign is the kind of concept (representation or theater by other names) that manages intensities by means of dualities (presence/absence). Yet the skin is not good and the sign bad. Or, to put it differently, certain set-ups are not better than others. They all need to be accounted for, and for that matter, "loved," Lyotard suggests. It is hard to say why exactly, he admits, except to reiterate the priority given to desire, and Lyotard's rejection of demands for clarity, as well as his recourse to specific words like "force," "anonymous," etc. Mostly, set-ups "drift," despite themselves, away from their moorings and this is what pure departures look like from the shores of Lyotard's theoretical masters (Marx, Freud, Saussure, etc.). Intensities are what matter here. And the more intense the thought of them, the more libidinal such thought is, which cannot be described as clear, but rather as tumultuous. The temptation for the reader is simply to plug into the circuit and enjoy the fireworks.

My focus is on "departures" from neutral signs (although Baudrillard certainly disabused semioticians of this notion). The so-called spinning bar already suggests the bar between the signifier and signified; yet Saussure's bar is not disjunctive, but like the cutting edge of the paper upon whose front and back of the sign's immaterial body is imagined. Lyotard (1993: 25) introduces two kinds of signs, intense and significational: "The intense sign which engenders the libidinal body abandons this vast Moebian skin to the significative sign, the singularity of a passage or a voyage of affects is herded, closed up into a communicable trace." The passage described here is from the singular to the structural based on distinctive oppositions, precise differences. How do the tensorial (libido, force, intensity) and the sign combine? How does a tensor sign sustain its singularity? For Lyotard, signs both replace (substitute) and call up the things for which they stand in for someone, and in this signification is dematerializing (additionally, such things are not outside of the semiotic sphere and are themselves signs, as the world is perfused, as Peirce would say, with signs). Yet dematerialization is not identical to deintensification (making something signify is still a force). It presents other opportunities for libidinal fluxes (1993: 44).

The bad names of semiotics are multiple in Lyotard's thought: nihilst, colonialist, simulator, preacher. These are all "games of de-presence" (such as endless decoding and the pursuit of meaning; recourse to lack and resignation) played against the intensities of tensors. By definition, tensors are extra-semiotic forces. They seem to belong to an extra-semiotic order. Yet this is only a lure. Lyotard (1993: 50) presents the option: if we quit signs, then we may enter the tensorial order. But if this path is taken, the semioticians can and will reconquer this order in due course. Lyotard's lesson is more advanced: "signs are not only terms, stages, set in relation and made explicit in a trail of conquest; that can also be, indissociably, singular and vain intensities in exodus" (1993: 50). So we must take great care, Lyotard says quoting Deleuze and Guattari, with our chosen "lines of flight." He is not relocating elsewhere in order to find or found a new kind of sign. Hence, I keep "departures" in scare quotes. Desire does not posit a post-semiotic utopia.

Philosopher John Protevi in *Political Physics* (2001: 59) provides a helpful analysis of Derrida's deployment of force and signification, separating force from oppositional couplings, and thus contrasting différ*a*nce with difference. Protevi directs us to a master metaphor: force and signification interweave in the signifying field of différ*a*nce, whose radical alterity is anathema to metaphysical oppositions (recall Baudrillard's critique of metaphysical elements of signification that gave the illusion of being rid of force). Active force is neither located in a meaning-giving act of a transcendental subject nor by metaphysical oppositions that apparently lack force or are presented as if the hierarchies they contain are natural. Differ*a*ntial signifying force undermines so-called stable and neutral oppositions (signifier/signified), as well as inherited hierarchies of the philosophical tradition (speech/writing; interiority/exteriority; sensible/intelligible) and in these ruptures release new possibilities in which grammatology overtakes semiology with the liberation of graphic substance and the neutralization of phonological substance (Derrida 1981: 27).

"The first thing to avoid, comrades, is pretending that we are situated elsewhere. We evacuate nothing, we stay in the same place, we occupy the terrain of signs," writes Lyotard (1993: 50). The signs that are interpreted, whose meaning-effects are plotted, these signs that "speak" to "you" move "us," libidinal economists, in different ways: set in motion, like a dancer (even an immobile pose, in silence). Lyotard loves his enemies, the semioticians, because he finds tension in their structures that "cover" over and dampen it. This is the "dissimulating" dimension of structure that "covers" an event and its affects. Semioticians, especially psychoanalysts, assign affects to one side of an opposition, the example that Lyotard gives is Freud's two principles of eros (unbinding) and thanatos (binding) and a cough from the case study of the hysteric Dora. Such affects work according to both sides, undecidably. Yet this is not merely both but something more, something beyond. Lyotard writes: "At the same time a sign which produces meaning through difference and opposition, and a sign producing intensity through force and singularity" (1993: 54). But Lyotard restricts his claim of prioritizing libidinal intensity—first it is the intensified throat of Dora, then the semiotic scaffolding is thrown up: "order matters little, what is, however, of great importance is the fact that this same symptom has inevitably two simultaneously possible receptions" (1993: 54). But not only two. At least two, and more. Carrying forward examples from Freudian case studies to the proper name as his primary example of a tensor's dissimulation, Lyotard points out that in accordance with philosophy of language, proper names are unlikely suspects because they refer singly, do not connote, and have no "intra-systemic equivalent" (1993: 55). Professor Flechsig is the psychotic Judge Schreber's doctor in Freud's case study of paranoia; for Lyotard, Flechsig is the "tensor par excellence" (1993: 58). The proper name is a spinning bar of disjunction that hovers around specific points on Schreber's body (mostly his anus), which Freud analyzed on the basis of the Judge's *Memoirs of My Nervous Illness*. Flechsig is more than either a protector or persecutor. The proper name Flechsig is the site of a localized and intense "vertigo," a term used multiple times by Lyotard to ward off meaning against the notion that such a tensor sign might be reduced to simple polysemia and an abundant meaningfulness. Freud traps Flechsig in a tightly scripted semiotic: Schreber's homosexual desire for his doctor. Whereas Lyotard allows Flechsig to activate many "incompossible pulsions" of love–hate between doctor and patient and release, not meanings, but serve as the extension of the eroticism (in the instance, anal) of the libidinal body of Schreber. The spinning bar spins "furiously" around Schreber's anus, which is the point of a vertiginous incandescence (1993: 60), and the name Flechsig is used by Schreber to anonymize his body; it is not incidental that in *Anti-Oedipus* Deleuze and Guattari turn at the outset to Judge Schreber's hallucinations of his solar anus to illustrate the productivity of desiring-machines (1977: 2). Thus, Lyotard accomplishes a dramatic turnaround: the proper name does not refer simply and straightforwardly, but is imbued with unpredictability, intensity, and anonymity, taking a place on a "scrap" of the libidinal band. And Deleuze agrees on this point, sharing another of Lyotard's examples from Pierre Klossowski's literary theory: the proper name

"Roberte" is an intensity "before designating or manifesting any person" (Deleuze 1990: 39–40; Lyotard 1993: 80). Further, Lyotard (1993: 55) mentions "Albertine" in Marcel Proust's *In Search of Lost Time* as a proper name that is, for the narrator, a "delirium." Guattari (2011a: 317–18) makes much the same point in his analysis of this novel and cites the breakdown of the "semiotization process" into a fractured, multiplication of fragments, and sensory implosions during an episode involving a kiss on the cheek.

Some of the features of Lyotard's spinning bar appear familiar within the philosophy of desire. In *Anti-Oedipus*, Deleuze and Guattari define the disjunctive synthesis of recording (1977: 75ff) in two ways: first, in terms of an either/or that restricts by differentiation (exclusion and restriction) one term for the sake of the other. This is, in agreement with Lyotard, its psychoanalytic use. Second, there is an immanent use. The shift here is not from one to both at once, but through an affirmation of more than both through the maintenance of disjunction: "either ... or ... or ..." (1977: 76). "Flechsig" is not both doctor and God (friend and enemy) at the same time but more than both, if by both is suggested a resolution of some sort (synthesis of contradiction; resolution of contraries). Instead, there is in an affirmative-disjunctive proliferation an opening out, a release of "ors," without restriction: "This is free disjunction: the differential positions persist in their entirety, they even take on a free quality, but they are all inhabited by a faceless and transpositional subject" (1977: 77). A subject who drifts (perhaps flies over) among the "ors." For Deleuze and Guattari, these flights are inscribed on the body-without-organs; for Lyotard, tensor signs bring into existence the libidinal skin, but with one exception—no inscription: "no assignable site of inscription. Only punctual incandescences, without instantiation" (1993: 74). And the intensities of the skin are weakened by exclusive disjunctions. The tensor sign is aligned with the immanent use of the disjunction. It is worth underlining this: the question for Deleuze, Guattari, and Lyotard is not meaning (search for a signifier and signified); it is productive use in a montage of forces.

Neither structuralism nor phenomenology is of much assistance in conceptualizing force, Derrida submits. Force is language's other, a "strange movement," without which it "would not be what it is" (1978: 27). Lyotard finds in the proper name an intensity that is occluded by its interpretation through invariability. This is not so far from Deleuze's route in *The Logic of Sense* in which Lewis Carroll's Alice loses her proper name, a "privileged indicator." Indeed, "the loss of the proper name is the adventure which is repeated throughout all Alice's adventures," claims Deleuze (1990: 3). It is the encounter with pure becoming that alters personal identity.

Let's review: tensor signs concentrate forces, focus intensities on scraps of the libidinal skin but without assignable sites of inscription (the skin retaining anonymity), generate singularities, induce vertigo, display ephemerality, incandesce, escape from and exceed positioning in oppositions. Lyotard eschews dialectics and presents no analysis of domination, sexual or otherwise. It is difficult to follow the "fault lines" between intelligent and intensive signs, Lyotard admits (1993: 76). Why? This is not an opposition that is immune from intermingling. Intelligence does not betray intensity. It is not a safe and permanent "shelter" from the barrage of pulses and fluxes, affirmations, and exorbitance, of displays of intensity and libidinal charges. Even the linguistic law of value can attract some charges of libidinal intensity in the dissimilar exchanges and comparisons based on similarities that it regulates. The key to grasping this Lyotardian insight is to observe that currency, the francs used to purchase bread and to compare with the other coinage and paper bills issued by the Swiss banks cited by Saussure, is not a "dead" body but possessed of a living body consisting of fluxes. Saussure did not consider money's body (i.e., the silver content of coins; the stamps on them, the arcana that excites the coin collector and displaces "face value") in his illustration of the theory of value. He sheltered ("covered") it from energetics. Likewise, he sheltered the chess pieces he evoked as rule-bound against the materiality of the pieces themselves. Lyotard's effort to define a tensor sign returns energy to the sign as a carrier of intense singularities that burn brightly and dissipate at the same time: "its syntactic 'ice' must be its incandescence" (Lyotard 1993: 79). Communication and incomparability are herein confused. Lyotard gives us an impossible situation: the meaningful communication of fluxes destroys them; the introduction of pulsional singularities into the law of value disables it. Near the end of *Libidinal Economy*, Lyotard reflects: "Must our fear of sign-systems, and therefore, our investment in them, be still so immense that we search for these pure positions?" (1993: 262) Yes. Conduct intensity, he says, invent nothing, release or follow tensors. Yes. And yes again.

My selective reading of Lyotard and the tensor sign emphasizes the importance of his creative approach to signs and his "love" and "fear" of linguistics. His introduction of intensity/energetics into structure is a precursor to the challenge to signs posed by affect, which appears not to want to engage in any kind of semiotic dissimulation. In concluding this section of the chapter, I want to reflect on Lyotard's earlier study of the linguistic sign in *Discourse, Figure* as a more straightforward case of his ingenuity regarding Saussure's work.

Thickening signs

The chapter titled "Linguistic Sign?" contains a remarkable intervention, the point of which is to find in signs a "thickness" that displays an "immediateness" grounded in both sensory and spatial dimensions. Lyotard's sense of a "thick space" (2011: 74) first exploits Saussure's (2011: 68) reflections on the symbol as possessing a "natural bond [or connection] between the signifier and the signified." The thickness in question is not re-embedded in the examples Saussure enumerated, namely, phonetic evolution, Latin derivation, the influence of "natural forces" in interjections, all of which have elements of "chance" (Bouissac 2010: 96). Lyotard considers the motivated symbol to be a "sensory signifier" that touches the body, through rhythm and corporeal experience, yet one of his examples is not a word but a visual one—a road sign indicating with two or more ridges that there is a "bump" ahead. Lyotard seeks a positive interpretation of symbolical relations not in the signifier–signified relationship (for the sign, in Saussure, is empty and has no interior, and its exterior is interior to the system of language) but in the extensionality of reference (primarily designation [deictics like "this tree," "here, not there," "now"], as I mentioned in passing earlier) in order to grasp that "the disposition of language induces upon words and between them rhythms that resonate with those provoked on our body by the thing discourse speaks of" (2011: 77). This argument is a precursor to the tensor sign as reference is itself a force tripped in the encounter with thickness and rhythmic space. Setting up his analysis in the zone between signs and things, then, Lyotard (2011: 83) claims against Derrida that there is something outside the text, in a "borderspace"—a "worldly type of space, plastic and atmospheric"—in which "motivation" exists.

The "thick sign" is a precursor to the tensor sign in Lyotard's critical philosophical and semiotic project. In another argument, Lyotard finds that there is an element of confusion in Saussure's discussion of the "well-defined object" of study (language), as opposed to speech, as such mixing itself poses questions of depth, distance, and designation. Lyotard claims that the designatum gets transposed with the signified and the latter displays an illusory depth at best (2011: 97). Geoff Bennington in his study *Lyotard, Writing the Event* (1988: 66) delivers the undoing: "'The linguist is condemned to the fate of all speakers: we cannot speak without tracing this distance between our discourse and its object': simply, the object of the discourse is language itself, which provokes that discourse to trace that same distance in the very discourse that argues against it."

Looking at Lyotard's reading of Saussure, he erects two axes at 90 degrees to one another—the vertical relation of signification between signifier and signified and the horizontal axis of value between signs; taken together these are proper mutual dependencies that may be expressed in this way: a sign requires signification and vice versa; yet both sign and signification are effects of differences with other signs, hence derivatives of value (Bouissac 2010: 100–1; Genosko 1994: 18–19). Thickness can also be imagined to exist in signification, as in the analogy of cutting a piece of paper ("one cannot cut the front without cutting the back ..."), in its "vertical depth," and in a kind of pseudo-exteriority between signs (Lyotard 2011: 93; 101). However, these are the "epistemological illusions" of how Saussure's *Course* was constructed around specific cuts, such as reference (exteriority) for the sake of the system (interiority) and external phenomena (history and materiality) for the sake of only what changes the internal system viewed synchronically. Therefore, Lyotard displaces the illusion of thickness in the sign onto the margins of discourse between words and things toward the domain of the nonlinguistic. Thickness poses quandaries of reference, introduces the sensory and visual, and reintroduces motivation: "the quasi-visible space introduced by designation is thus absolutely heterogeneous to the unmotivated function of linguistic space" (Readings 1991: 16).

For semioticians, it is hotly contested whether a deictic indicator needs an actual referent, and if one argues that it is does not, then Lyotard's position becomes less attractive. However, if signification is in some respect contaminated by designation in the case of deictic terms, such that the signified–referent distinction is also troubled, then Lyotard's insight into instabilities within Saussure's internal linguistics must admit of a "distance," a sensory field, a depth and thickness. This is how he aligns the figural with designation. But, a semiotician like Eco maintains that deictics do not require "testable referents" (1976: 119) in considering examples of both words and gestures ("this-that" and pointing fingers). The physical nonverbal gesture of pointing displays some of the features that Lyotard identifies (stress on closeness or distance of the object indicated in the finger's case, and dynamism, expressivity) yet within a semiotic framework. However, in Eco's (1976: 116) comments on "this-that" and "here-there" there may be nothing nonlinguistic ("even if the presupposed event or thing does not exist and never has existed") at stake in understanding these terms. For Eco, "our theory excludes physical connection with the referent and considers closeness as a signified content" (1976: 116). This is diametrically opposed to Lyotard's position where distance, whatever it may be, is spatialized, not signified. Now, it may be that Eco is simply unknowingly rehearsing what Lyotard is arguing about the linguist's fate of taking language (or, specifically, the sign) as the referent of his own discourse, even though his discourse excludes the referent; although Eco tries to solve this problem by building a theory of codes in which designation conveys not a referent but "cultural content," it is naïve (a "referential fallacy," for him) to consider the matter otherwise! Lyotard is not one of these naïve persons, and both thinkers seem to agree that Saussure treated the signified somewhat unclearly, anyway. For Lyotard (2011: 98), the signified seems to be modeled on a referent that is excluded. Lyotard's position is extreme: "there is no signified, except as mirage." As Bennington (1988: 66 and 69) rightly underlines, this is neither a correction and restoration of an error in Saussure's thought nor a reintroduction of referentiality, but an opportunity to introduce thickness into the heart of structural linguistics, that is, depth and seeing, into an otherwise readable, flat space based on the axis of value. But Lyotard's theory does not end with this distinction in which one pole (depth and seeing) is valorized, as he then proposes to find discourse within the most figural settings. My goal is not, however, a full explication of Lyotard's system. Suffice to say the link proposed between the arguments about figurality and energetics conjoin in the deployment of forces within structural linguistics in the use of proper names and designation as privileged examples that run against the grain of received wisdom in philosophy of language and much semiotics. Lyotard does not require a real object as referent to verify his theory of the border space of vision within discourse; rather, therein objects appear in a variety of guises and varying degrees of reality-testing such as actions that prove and names that may be communicated, and fantasy as in object-seeking that may be realized in images such as paintings (Lyotard 2011: 280–1).

The critical semiotic question is this: instead of quitting semiotics proper and investigating signs that glow with strange and competing anonymous intensities, we accept signs themselves perfused with such intensities as pathways to an analysis of affects and the becomings that they induce. How, then, does affect intersect with signification?

Affect theory

The answer, at least at first, is awkwardly. One of the sources of this is Brian Massumi's book, *Parables for the Virtual*, that announces an ambitious plan to refresh cultural theory by regaining the bracketed middle of movement/sensation between bodies and change. Cultural theory (including semiology) lacks processual awareness and is too grid-like in the way it pins down bodies in its mode of apprehension, especially in the identification of local manifestations and resistances. Massumi wants to give matter and motion, potentiality, processuality, dynamism, intensity, and becoming back to bodies: "rethink body, subjectivity, and social change in terms of movement, affect, force, and violence—before code, text, and signification" (2002: 66). The requirements are many: abandon linguistic coding; put movement before position; regain Peirce's concern with continuity (Massumi himself regains his own reference to Peirce on continuity in his later book *Semblance and Event* [2011: 88], noting the virtual surfeit of feelings shimmering around any actual feeling); reject preexisting interpretive grids that arrest motion and are not abstract enough. Such grids bracket becomings. Structures can smother events. Language can dampen intensities (affects) by codes "in which all eventual permutations are prefigured in a self-consistent set of invariant generative rules" (2002: 27). Process must come before signification and semiogenesis. The general problem Massumi elucidates is the effort to put movement back into code: "after all is signified and sited, there is the nagging problem of how to add movement back into the picture. But adding movement to stasis is about as easy as multiplying a number by zero and getting a positive product" (Massumi 2002: 3). Despite all of the richness and nuance in the development and execution of this project, the absence of a sustained reflection on the contributions of post-structuralism and the efforts, like those of Lyotard and Derrida, to think intensity, force, and signification, leads to a further problem: that movement may not have been completely missing in the first place from codes, structures, and discourse. Of course, the degree to which forces were stirring in the Saussurean project and in the work of its major interpreters remains a matter of contention, but at the very least recourse to Lacan's famous formulation of S/s—signifier over [bar] signified, the two orders "separated initially by a barrier resisting signification" (1977: 149), not to mention his observation about the "sliding" of the signified under the signifier (1977: 154)—would provide examples of how certain versions of the sign marshall force and movement. And Barthes (1967: 49), for instance, reaffirms Lyotard's observation of the illusion of depth in the sign, adding that Lacan attributed a "value" of repression to the bar; Anthony Wilden (1968: 238) refers to Lacan's "inverted version" of Saussure's sign and expertly traces down the variations that this version undergoes. Refreshing the linguistic turn, if undertaken via post-structuralism, does not lead to what some have figured as a "stalemate" (Brown and Tucker 2010: 234) in the form of a semiotic stasis. The continuing importance of Peirce has already been noted, although he stands outside of and is often invoked as an antidote to semiology, and recurs in Massumi's myriad deployments of pragmatist thought and the posing of semiotic questions. As Guattari once put it, he wanted to reinforce the distinction between semiology and semiotics, often favoring the latter because it did not "depend on linguistics" (2011a: 22) and survived repeated attempts by linguistics to annex semiotics.

Lyotard's brand of post-structuralism and critical semiotic thought answers the charge that affect theorists have made against semiology with a rousing flood of libido into signs. The absence of a reflection on power has been noted in passing above, but it is worthwhile reflecting further on this issue in relation to Guattari's approach. He asked after how power "crystallized" in signification—encoded through dominant grammars and repetitive structures such as refrains that reinforce standards of proper speech adapted to the needs of capitalistic communication; a semiotic sorting operation is undertaken to select the conformist elements from the intensities that do cross language in creative flights, but also to manufacture them whenever possible, and to stifle signifying potentialities and impose coherent abstractions so that any persistent asperities are smoothed over. A good example of variation within language may be seen in the proliferation of themed Internet memes, with their humorous agrammaticality, unattributed authorship, and stylistic consistencies, not to mention irreverence and irrelevance (Davison 2012). These carefully executed visual and textual knots contribute to the promotion of a machinic slang of emoticons, deliberate misspellings, and acronyms (and many initialisms). The agrammaticality of Internet memes (i.e., LOLcats) is not simply a contingent feature, a feature of speech, but places grammaticality in "continuous variation," conjoined with amateur photography and favored typography, with identifiable procedures yet without authorial attribution (Deleuze and Guattari 1987: 93–9).

Further, Guattari (2011a: 32) asked how signifying statements "crystallize" intensities. Multiple intensities, including both affective and libidinal, crystallize in signifying discursivity. However, this crystallization is not limited to the intersecting axes of signification and value in the Saussurean system, the very stultifying differences that affect theorists resist, while in addition flagging the impingements of numerous other structural delimitations, and which Lyotard worked hard to complicate with tensorial flows. As Lone Bertelsen and Andrew Murphie (2010: 153) put it, following Guattari, "signs, in their affective dimension, burst out of 'strictly linguistic axiomatics' ... [and] exhibit a 'non-discursive aspect', one that acts beyond the constraints of discourse." What is suggested here is that the pathic apprehension of otherwise overcoded significations may release, admit, or transmit under- and other decodings, that is, nondiscursive intensities.

As I discussed in the first chapter, Guattari presented the relationship between discursive (space–time–energy coordinates, syntagms) and nondiscursive (pathic, affective, intensive) fields in less oppositional than mutually imbricating terms, with the proviso that it is characterized by two tensions. Small supplies of discourse exist in the pathic field, and they are used to discern, modulate speeds, and self-actualize. So, pathic life needs a certain amount of discourse, but because discourse wants to bracket pathic apprehension in order to pursue its own aims, namely, denotation and locatable exo-references, pathic experience must marshall forces to divert discourse from these ends, finding new roles for it. But discourse may be said to, paradoxically Guattari (1995: 26) admits, rest on pathic experience, as all subjectification has a pathic foundation. Instead of binarizing, then, Guattari creates a mutual imbrication. These are concepts that inform the distinction between signs and affects. Thus the semiotic is not a separate paradigm, but supplies the theorization of affect with the means to express encounters between bodies and communicate openness to the virtual potential of relations, even though certain signs tend toward static matrices that are good at arresting, containing, and decomplexifying. It does not follow that the semiotic rests upon a foundation of affect, a presocial and prelinguistic bodily capacity to affect and be affected. There is no genesis, no early state to be regained and reactivated. Rather, the first curve in the "turn" is to rethink encounters, movements, and intensities before signification, code, and discourse. The "asociality" of intensity is more or less agreed upon, and chicken and egg questions are eventually abandoned (Clough 2007: 2; Massumi 2002: 30). Refreshing the conceptual vocabulary to complete the affective "turn" means mixing it up in-between semiotic and affective elements. This is not quite what Lyotard insisted on, producing meaning and intensity at the same time, but instead, moving in-between them, not for the sake of the joys of aporia, but for that of pathically apprehending change and becoming in their own right.

The fate of energetics

While Guattari in his *Schizoanalytic Cartographies* (2013: 51) wrote of "quit[ting] the terrain of signifying interpretation" for the alternative he proposed with regard to the study of the unconscious that focused on affects and technoscientific effects, it is evident that he didn't altogether leave semiotics in the process. Rather, he wanted to "quit" existing tendencies to create binaries and dualities, especially those arising from the assimilation of flows into forms or, the capture of matter by an imposed form that leeches out the specificities from matter in the name of a "single economy of energy." Both Freudian and Marxian theory suffered from what Guattari dubbed an "infrastructure complex" built upon the circumscription of givens by univocal energetics, such "monotheisms" of Capital and Libido. Guattari's solution is to pluralize energies and thus "quit" univocal versions of them. Guattari complained that Lacan substituted the signifier for libido and as a corrective he wanted to keep energies in the plural together with modes of semiosis. What he imagined was not a science that would detect a universal energy in a specific substrate, but an energetico-semiotics that could be diagrammed in terms of the interdomain relations of his quadripartite framework (phyla and fluxes; universes and territories) and in this way he could describe the levels of energy belonging to the entities arising from the interactions of four domains (existential territorial refuges; alterities of incorporeal universes; machinic phyla; material flows; effects between flows and phyla and affects between territories and universes). It would be possible, he thought, to isolate and describe the quanta (minimal amounts or "elementary" packets) of energy arising from interdomain relations. For our purposes here, suffice to say that Guattari's highly esoteric and inventive mixture of science and magical properties poses the solution of the energy-signification relation with recourse to tensors that connect and mediate forces across ontological domains.

Guattari's borrowing of terminology here is less about separating meaning from intensity as in Lyotard's use of the term, and more about describing how entities pass between domains and in the process are transformed, but not beyond the "laws"—well, the constitutive definitions of how entities are characterized and ways they are organized —that they supposedly obey, the vectors along which they are projected, and the charges they carry. Tensors in Guattari's usage also actualize and virtualize, that is, manifest a potential in the actual or potentialize a given by means of development. Unlike Lyotard, however, they are explicitly deployed to dissolve the intensity-signification problem; and like Lyotard, Guattari introduced elements of movement and change in multiples, both quantitative and qualitative. Although Guattari retains some of the conceptual language of semiotics (expression/content; synchronic/diachronic), his piling up of concepts and definitions around them, augmented by diagrams, makes semiotics vertiginous. Simpkins (2001: 206) may have been right in *Literary Semiotics* in pointing out that Lyotard engaged in a critique of signs that was written in a "caricatural register" by exaggerating the closed, finite nature of sign-play (i.e., Lacan's version of the entire signifying chain as the mediating device of any possible reference) and the lack of progress and action in closed systems, by rejecting this effort as "disturbing," he leaves us in an unenviable and perhaps untenable position of trying to inject some movement into a largely static structure. Perhaps Massumi was not altogether correct when he used the idea of multiplying by zero as a way of characterizing the intractable problem of getting change back into structure, a place where nothing allegedly ever happens. Instead, if we consider Lyotard's example, recourse to division by zero is a more apt way of indicating that the sign's becoming intense is also a case of its becoming more indeterminate and more open, that is, undefined. Although for some critics, and at times for Lyotard himself in accounts of his own work, this produces an error message, just as a calculator generates such a message when division by zero is attempted, the search for an intense sign remains a productive philosophical conundrum for critical semiotics.

6

A Toolbox for Critical Semiotics

In this final chapter, I will turn to a select few concepts that comprise a toolbox for critical semiotic theory and in the process a number of the themes from previous chapters will return and intertwine. Guattari (1996a: 173) once remarked that he "had the privilege of seeing Michel Foucault take up [his] suggestion…that concepts were after all nothing but tools and that theories were equivalent to the boxes that contained them." Of course, in this constitution of tools it is understood that they may be hacked and then put to counterpurposes for which they were not originally designed, a practice that may introduce an acceptable degree of fragmentation and the production of odd theoretical bedfellows. Embracing these risks and advantages, I turn first to the vivid material signifier of Fredric Jameson's study *Postmodernism or, The Cultural Logic of Late Capitalism* (1991), with the proviso that it will be stripped of its lingering semiological trappings, although it is defined against Saussure's postulate of negative difference by depriving itself of it in an effort to characterize the "cultural dominant" of postmodernity. This too is altered in order to reveal its relationship with the affective, energetic, and figural problematics of Lyotard and Deleuze. The key passage is through Saussure and Lacan and an exit point is indicated by Deleuze and Guattari.

The second task in this chapter is to revisit the floating signifier posited by Claude Lévi-Strauss in his *Introduction to the Work of Marcel Mauss* (1987). Although a trope of flotation is favored by Baudrillard among others, it is rather its status as a "semantic function whose role is to enable symbolic thinking to operate despite the contradiction inherent in it" (1987: 63) that is at issue. Managing the fit between a "signifier-surfeit" and the pool of available signifieds requires a special "supplementary ration" that is a semiological translation of how symbolic thinking manages exchange. For Lévi-Strauss, the existence of a zero symbol in a "pure state" is the solution that corresponds to Mauss's use of mana-type concepts bearing "secret forces" (that which animates the *taonga* [objects of value], the inanimate spirit of *hau* among the Maori people; Mauss 1990: 14–15) in the exchange of gifts: "a sign marking the necessity of a supplementary symbolic content over and above that which the signified already contains, which can be any value at all" (1987: 64). Reviewing a number of philosophical reflections on the floating signifier and signified leads to the consideration of the value of zero as a critical semiotic conceptual tool. In the end, I will revisit the problem of searching for a positive outcome using zero despite, as it is said, *ex nihilo, nihil fit*.

Shiny signifiers

Jameson (1991: 6) gives to the postmodern the figure of a "force field," what appears to be an entirely fictional entity, crossed by impulses whose dominant features, primarily their traces and protentions, he wants to capture. The specific exposition that will concern me in this chapter is Jameson's recourse to Lacan's theory of psychosis in a semiological frame as a way of reading the schizo-syntax of contemporary art's temporality. Despite his observation that postmodern art displays, with certain limitations, the "waning of affect," which seems untimely given the subsequent turn to affect in contemporary theory, and regardless of his penchant for characterizations of an "ideal schizophrenic's" perception, fascinated in and by anything in the present, Jameson's contributions to understanding the brokenness of sign relations in postmodernity are an indispensable resource for critical semiotics. Conclusions about the role of affect in Jameson's thought should not be jumped to as affect will return in a strange form. However, the aforementioned waning is accompanied by the disappearance of the centered, monadic self/subject, and the waxing of the subjectless subjectivity, not only crossed, but buoyed by intensities: "free-floating and impersonal and... dominated by a peculiar kind of euphoria" (1991: 16). Moreover, if for Lyotard the waning of metanarratives (of modernity) marked the postmodern turn, Jameson adds to the theses about affect and the subject the further waning of historicity and the waxing of the present, though a peculiar present it is because it fails at representing itself and shines like a "glossy mirage" (1991: 21). The post-subject of postmodernity, then, in Jameson's estimation, is as Doug Kellner (1989: 29) once noted exposing the paradox of a Marxist critic "dissolving the subject into schizophrenic pieces hardly seems the most productive route for a new radical politics [...] but is rather a challenge to overcome and surmount." Whether the sign in crisis is reparable, "breaking up in mid-air" and sprinkling "pure [material] signifiers" that Jameson describes elsewhere (1984: 201), seems a foregone conclusion if surmounting is a belief, utopian or otherwise, in restoration. From the "junk" of postmodern signifying dissolution something will rise or, perhaps better, the sign's crack-up, splintering, and flaking supplies the semiotic dust plateau with twinkling fragments.

Jameson's analysis of the breakdown of signs deploys vividness to describe the suspended present of the postmodern condition in which intensification occurs that renders asunder a signifier from the others with which it is interrelated through negative differences. This breaking-apart both isolates and vivifies. These related effects are tied to the role that tense plays in Jameson's analysis of how meaning is generated semiologically. Once the "simultaneous presence" of other "interdependent terms" noted by Saussure (2011: 114) is observed to fall away, and the materiality of a signifier in isolation emerges, a further principle of structural linguistics is violated: it is not the materiality of the linguistic signifier that interests Saussure in determining its value but, rather, its differences within a closed system (as we have seen, value should not be confused with the "tangible element" or by analogy the metal in the coin whose value is determined by the amount stamped on it within the currency and how it is exchanged for goods rather than the equality of value between the metal and the stamped amount). This is of course not at all sensible to a collector of coins whose goal is to exploit this material differential against the immaterial, putting the support ahead of the negative, incorporeal differences. The violation of this most "basic" principle as Saussure refers to it sets up the contrasting analysis and puts into relief the severity of the situation and the apparent necessity of going all the way over to psychosis to describe it. Although Lyotard issued a challenge to this (via written signs and letters) in the concept of the figure that refocuses attention both on the means of writing and the visual results, suffice to point out that Jameson's aesthetic preoccupations illustrate well this post-meaning effect or the incapacity of an independent signifier to project a meaning-effect. Independence is anathema to the interdependence of the structural signifier. Separate, stranded, a snapped chain, and purely material: Jameson's attention to the hum and glow of this signifier is accompanied by the temporal observation of "pure and unrelated presents in time" that apparently describes schizophrenic experience. Thus, the two breakdowns occur together (signifying chain and life experience) and in the latter case time loses its capacity to unite and solidify past, present, and future: "the breakdown of temporality suddenly releases this present of time from all the activities and intentionalities that might focus it [...] thereby isolated, that present suddenly engulfs the subject with undescribable vividness" (1991: 27). If affect had previously waned in Jameson's estimation, it returns in a way that is "overwhelming"; now, it is a "mysterious charge" with either euphoric ("hallucinatory") or debilitating ("anxiety") implications for the subject. The signifier's energy is obscure. Considered in aesthetic terms, Jameson's sensitivity to "vivid perception" is discussed through poetic examples (textual practices), and he cites collage as an example of a mode of perception and apprehension required to view postmodern works. The lesson is obvious: signifiers conduct energy only when they fail.

After the collapse of lived time, and the emergence of a shattered time that consists of isolated fragments, so many nows realer than real, which correspond to presents like videograms that may be cut from video recordings, bring back the issue of tense. Shifts in tense can be used to mark decisive moments in narratives, and the emergence of isolated presents illustrates this nicely in Jameson's own theory; however, *vividness* is an interesting choice of term as it refers to varying degrees of events expressed in future tenses in conditional forms (some more, others less vivid). Here, Jameson transposes this lesson to describe presents-most-vivid. Jameson stays within the purview of language here and accepts a semiological explanation of schizophrenia without putting up any resistance; whether he can be considered Lacanian in any meaningful sense is in doubt by the absence of an explicit consideration of the way in which access to the Symbolic order is prevented by the foreclosure of a special signifier (Name-Of-The-Father) and results in the substitution of other broken signifiers that pile up in the hole in the signifying chain (Lacan 1977: 215). He does remain within the Freud–Lacan understanding of language as the key to psychosis, bearing upon the Schreber case in particular, but the fate of the paternal metaphor does not play a role. Nevertheless, Jameson's recourse to vividness expresses that the signifier has become an enigma for schizophrenic experience, and in its isolation throws up a roadblock to re-integration into the signifying chain, not to mention being pinned onto a signified, or fixed by a code. Within schizophrenic experience defined in this manner messages are constructed, but are communicated perhaps in a clang association of sounds that are wielded like things, or glowing things animating the rusty ruins are treated like concepts. It would be prudent to remind ourselves that the escape from language as the central explanatory model in psychoanalysis was proposed by Guattari, toward a mixed semiotic that includes a-signifying elements as well, not far removed from the vivid signifiers Jameson proposed. But why can't these isolated signifiers be added together? They cannot be semiologically formed, which is different from asserting that they may assembled (i.e., in a delirious refrain) in a nonbinary way that does not require full formation and does not conform to a single set of systemic rules of the game. Guattari's rejection of the emphasis that Lacan put on the paternal metaphor in his theory of psychosis is positively anti-oedipal and anti-semiological, but this does not influence Jameson's theoretical construction.

What kind of resource does Jameson provide? The materiality of the signifier moves toward Lyotard's (2011: 212) contention that the graphic line possesses a "figural power [that] can only break out, like a scandal." Not, then, a breakdown, but a breakthrough. Vividness and glow reorient attention to the sensory and coax the negative and differential into the background. Both Jameson and Lyotard interject the body into the semiotic scene that the latter describes as consisting of two spaces, graphical and figural: the body of the reader/viewer is affected by the material signifier and the plastic values it possesses, beside the issue of signification and value. Where there is no confusion for Saussure on this point, Lyotard introduces a good measure of it, specifically noting changes in comportment before letters with multiple straight-line segments stacked together (N and A). Citing the ancient pre-Socratics, Lyotard concludes: "The moment one endeavors to transform the world into a text, one is tempted to smuggle a little of the world back into the text" (2011: 207). Jameson provides little opportunity for adjustment of the subject engulfed by material signifiers as the "space of praxis" collapses. But this is the point: he acknowledges that the space of an encounter is at issue, but only to provide two extremes (anxiety or euphoria) and conjecture about perceiving the positive relationships between differences ("radical and random"), in video art and cinema. Already Jameson has shifted away from the semiological, what Lyotard characterizes as the "informational space" of communication, into the figural as the viewer changes focus from one element to all elements and "to rise somehow to a level at which the vivid perception of radical difference is in and of itself a new mode of grasping what used to be called relationship" (1991: 31). Loss and gain: Lyotard's words for the uneasy stops and starts of the energetics at play in line and figure relations. Jameson, too, thinks in terms of energy, recall above the "force field," but he is doubtful about whether viewers are up the challenges of postmodern art and thus may not be able to navigate the field. It all depends on where the energy is deposited: in the line of a great painter like Paul Klee, or in the vividness of the stranded material signifiers. Lyotard, for his part, also needs a temporal dimension in his analysis: the eye slows down because the speeds of the graphic carry it forward at too great a pace ("impelling thought") to recognize the eruption of plasticity's potential. Discourse deprives us of the pleasures of slow theory.

Signification overrides sensory apprehension. Legibility tends to erase plasticity. These are affects in the Deleuzean-Spinozist (Deleuze 1988: 125) sense of the slownesses and speeds of a body that may be composed in the midst of the confusion between graphic and plastic, letter and line. This makes Lyotard into a good Deleuzean concerned with affective capacities of bodies and the experimentation required to find their thresholds and theorize the constraints that hold them together. This is something that, in a hushed tone, Jameson may be reaching for in his recourse to the older term "relationship." But he is not a philosopher of desire.

As a final remark on this conceptual tool, it may be sobering to remind ourselves about Lévi-Strauss's reflection in *Structural Anthropology* concerning the arbitrariness of the sign. He claims that arbitrariness is "provisional" once linguistics is embedded in the study of social symbolism. In other words, arbitrariness in a code such as the color of traffic signals, even if reversed so that red comes to mean go and green means stop, remains arbitrary yet something occurs beyond sensory impact and oppositional definitions. He refers to the "inherent value" and "independent content" (1963: 94) of a sign that modulates its meaning by bringing into play the constraints associated with traditional, historically embedded, symbolism, in the manipulation of which "complete freedom" is impossible. In other words, Lévi-Strauss is not satisfied with recourse to the sensory experience of intrinsic properties and the sign's arbitrariness, yet brings embodied cognition into play in decoding while also insisting on independent factors with the power to modulate signification and reorganize sensory life. His recourse is not to painting and poetry, but to Marx's musings on gold and silver, whose differences in the ways they reflect light make them aesthetically interesting choices, but they are also woven into a web of other symbolic social relationships. What are the implications for critical semiotic theory of positing an "independent content" that is supplementary?

Floating signifiers

My next stop on this tour of the theoretical tool shed is the infamous "floating signifier." Jeffrey Mehlmann traces the concept from Lévi-Strauss to Lacan and provides a good summary of the tension ("paradox of knowledge") that the concept is supposed to relieve: "Whereas the linguistic theory of totality (of meaning) must have come into existence (as structure) all at once, that which we know has been acquired progressively" (1972: 23). The dissymmetry between an overabundance of signifiers and not enough signifieds requires a versatile and abstract supplement with the following properties: it must have an undetermined quantity; be devoid of meaning, yet able to attract any meaning; and accept the task of bridging the sign's strata when required. Lévi-Strauss's conversion of the principle outlined by Mauss of the spiritual power of the gift, which enjoys a certain amount of autonomy in its own right, suggests that the floating signifier is supplementary to signification yet not mapped in advance. It exercises power by, in specific instances, ensuring the working of a code.

Portuguese philosopher José Gil devotes himself to floating signifiers at length in his *Metamorphoses of the Body*, and explains that flotation refers primarily to the lack of an anchor, where there is a signification but it has no assigned meaning. Gil emphasizes the functionality of floating signifiers for primitive thought—they not only allow symbolic thought to operate but provide a means to transition between codes. Such signifiers are distributed into "semantically disordered zones" (1990: 94)—they are creatures of boundaries—and bring their energy to bear on individuals, humans, and things. For Gil, "the body is the exchanger of codes. It is the body—and its energies—that the floating signifier refers to" (1990: 95). By facilitating code exchange, the floating signifier enables signification: "if it signifies nothing, then it allows signification to come about" (1990: 98). It is often a shaman's task to bring sign and object and signifier and signified into contact and to provide an efficacious means of expression (a recitation or song) to the afflicted as an anchor.

Gil's examples are focused on bodies, and he begins with the art of mime and points out the suspenseful relationship between the signifier and signified in performance: "The signs produced by a mime unsettle us because they cannot be detached from the signified inscribed in the body itself" (1990: 108). He accounts for the array of artful and affectively rich microgestures produced by mimes that lack meaning in themselves and conspire together in a cluster of forces to reach the bodies of spectators and impact them directly. A mime's body pushes toward the limits of a metalanguage without speech, yet the body does not speak: "On its own the body signifies nothing, says nothing. It always speaks only the language of the other (codes) that come and inscribe themselves on it" (1990: 98). The body is the reference point in this phenomenological adjustment of its own turn to language, and Gil would agree with Lyotard on this point, so as to include the reinsertion of the sensuous and the material into signification, even if as Lyotard has claimed, it is most evident on the borders of discourse.

Floating signifieds

Deleuze's lecture "Eighth Series of Structure" in *The Logic of Sense* adds another wrinkle to the Lévi-Straussian conception of flotation. Deleuze is interested in two heterogeneous series, as he calls them, of the signifier and the signified, and the relationship between them he names the "Robinson's [after Crusoe] paradox." Representing Lévi-Strauss's explanation of the floating signifier, much in the manner of Mehlmann whose work was cited earlier, Deleuze considers the two series of signifier and signified to be antinomous and in "eternal disequilibrium" (1990: 48). Antinomy advances logically the anthropologist's use of "inadequacy"; regardless, Deleuze views the gap between the series, the first marked by excess and the second by lack, as an ontological problem of whole and part. Whereas the series of the signifier is that of a "preliminary totality," the series of the signified is that of a "produced totality," arrived at progressively. The shipwrecked Robinson can establish the rules of society all at once (signifier), even though his conquest of the island will be partial and his progress lurches from one thing to another (signified). Robinson must avoid the errors, argues Deleuze, of reformism (or technocracy, which amounts to the same thing) and totalitarianism; the former attempts to partialize the social totality, while the latter tries to totalize the partiality.

The virtue of Deleuze's account of the two series is that he views them as "two sides of the same thing," which he expresses abstractly:

> SIGNIFYING: excess ... floating signifier ... empty square without occupant
>
> SIGNIFIED: lack ... floating signified ... occupant without a place

The gap between the series is filled by either floating signifiers or signifieds as long as the empty square theory, based on either anthropological concepts like *mana* or psychoanalytic concepts like *ça* [it], which are empty and void of meaning, can take on when required any meaning and fill the gap momentarily in an interchange without equilibrium. Deleuze's quarry is how to define structure: it requires at minimum two series, defined by mutual relations, to which correspond singularities that may be joined by special differentiators (the technical term Deleuze deploys to talk about solving inter-series communication is "differenciation") that circulate among the series either as excess or as lack, hence, the requirement for floating signifieds, too. Whether Deleuze ends up falling back on the signifier, and really doesn't empower the signified, should concern the critical semiotician. If it is the case that the empty square of the signifying series is more important than the signified series' lack—the last line is "we can conclude that there is no structure without the empty square, which makes everything function" (1990: 51)—does Deleuze fail to extract himself from the kind of signifier fascination that he pillories elsewhere? Or, do we simply have to wait for him to get back to demolishing the signifier later along the trail of his writings? Patience is in order. The deployment of flotation is already in the service of deterritorializing the linguistic sign and understanding what encounters with signs give rise to (interpretation), as well as his deviations from Peirce and Hjelmslev. The question may be restated pithily: whither Lacan?

The remarkable reflection contained in the essay "Function and Field of Speech and Language" by Lacan is a meditation on the unconscious. The laws and structures of which, especially their symbolic effects, both the anthropologist and psychoanalyst share an intimate interest in. Constraints on freedom of choice in marriage, and on the drift toward incest, are approached through literary examples that set the stage for anthropological discoveries, which in turn are expressed in the shared language of structural linguistics, with an inflection toward the algorithmic combination of minimal units called mathemes. In this context, Lacan revisits Lévi-Strauss:

> Identified with the sacred *hau* or with the omnipresent *mana*, the inviolable Debt is the guarantee that the voyage on which wives and goods are embarked will bring back to their point of departure in never-failing cycle other women and other goods, all carrying an identical entity: what Lévi-Strauss calls a "zero-symbol," thus reducing the power of Speech to the form of an algebraic sign. (Lacan 1977: 68)

Lacan directs his redefinition of the floating signifier to his penchant for algebra, and his attachment to zero is complicated because it is empty yet functional to the extent that it reveals, like a shaman, that speech (the psychoanalytic talking cure) functions because it is empty and can take on any content, including algebra, thus bridging the gap between the signifier and signified and expressing the truth of the subject. Mikkel Borch-Jacobsen astutely (1991: 161–2) calls Lacan's mathemes "magical formulas," emphasizing not the rationality of their deployment but their fictionality and obscurity. If they truly "display their own senselessness," as Borch-Jacobsen insists, we can say that zero is puzzling. Deploying it in an analytic situation, then, is like multiplying any number by zero, which rehearses Massumi's procedure (he, too, invokes the "zero-point," 2002: 3), such that one doesn't generate more Lacanian analysts as positive results, but none at all. This has evidently not happened as Lacanian analysts abound internationally. So, after all, is nothing times nothing something? In desiring to occupy the empty square of the analyst, the analysand forgoes the occupation of the empty square of the absolute master and gives up speech for the studied abstentions of analyst and member of one of many Lacanian splinter organizations.

Wilden (1968: 128, n. 98) reminds us of the zero symbol as derivative of the zero-phoneme (i.e., a voiceless vowel) that doesn't have an assignable positive value phonetically or differentially, except in the global sense noted below; Saussure's examples of zero-signs are from Czech (2011: 86–7). While it makes no sense to claim that adding nothing to the phonemic inventory increases it, it may be that qualifying as zero is not so difficult. By the time the zero symbol filters down through the linguistic tradition, through anthropology and psychoanalysis and into semiology, we can find Barthes (1967: 77) qualifying it as a "significant absence" rather than a total absence. In providing a multidisciplinary catalog of zero concepts, Barthes looks for applications of Roman Jakobson's (1971: 431) definition: "zero-phoneme ... is opposed to all other French phonemes by the absence of both distinctive features and of a constant sound characteristic. On the other hand, the zero-phoneme is opposed to the absence of any phonemes whatsoever." What makes this zero "significant" is that stands in opposition to the total absence of signification. This so-called significance, while embedded in a strict discussion of "privative oppositions," is a double absence defined by opposition to all other phonemes and the absence of any phonemes. Here the question of whether there is a positive product of such absences proves vexing: in some circumstances, two negatives can make a positive ("not not"), but in this case the positive effect of the zero-sign must be understood as pure potential—another energy field—

that has the power to attract change in the form of meaning. This is a virtual attribute: it assists in the creation of actual meanings by providing the conditions for them to occur on contingent bases. Potentialities may persist in excess without being actualized; they may also emerge and then recede. This is a different sort of virtuality than that of associative relations as Saussure described them in terms of lack of order, circumstance, and of an unpredictable quantity: "A word can always evoke everything that can be associated with it in one way or another" (2011: 126). The Deleuzean virtuality at issue has the following attributes:

i. the virtual exists to be actualized;
ii. it is real even before it is actualized;
iii. actualization is not representation;
iv. the virtual possesses a creative capacity. (Boundas 1996)

The "despotic signifier," as Deleuze and Guattari (1977: 208–9) called it, gladdened the hearts of the acolytes of structuralism and gathers around itself and stingily distributes all of the flows, forcefully imposing the rules of difference that allow language to overcode the subject: "one" (master) signifier does what any other signifier could do, represent the subject for all the other signifiers. There is no joy in this. However, instead of overcoding all fluxes, the virtual sign manages flows, some of which actualize themselves minimally (as parts) and contingently, but mostly flow through. This might be a "linguistics of flow" (Dosse 2012:136) that would not be arrested by actualizations but persist as it retains an element of contingency and unpredictability, if language was still at the center of signification. Occupying the empty square does not close down the flow, and language is not the only occupant. Flows are not formed into discrete packets but their continuities are modified by the introduction of different pressures, changes in velocity, etc. Consider this in the inverse. Flows of various kinds can actualize themselves as floating signifiers or signifieds, because they are potentially signs, among other actualizations, and they bring varying degrees of intensity to the process, some countereffective (structural linguistic), and some hypertensile (tensors, affects), the others receding into latency. Although the problem of affect theory is how to get change and movement into structure, the problem here is how to recognize that structure is potential for change and movement.

Zero plus

Might Lyotard assist in this project of understanding zero? Not exactly, as he discredits "the great Zero" throughout *Libidinal Economy*, under which he subsumes all metanarratives, universal histories and theories, and ontologies of desire, deferrals, orders of simulacra, and detours. As Stuart Sim puts it in *The Lyotard Dictionary* entry on this subject, "the great Zero assumes a master signifier which endows everything with meaning... but of course it can't explain itself; it has to be accepted as a given whose value is transcendent" (2011: 91). All such theories, representations, and points-of-view are considered to be empty, nothing. And does this dismissal include Lyotard's own concept? His approach is allusive. He views Nothing as libidinal, enjoyable, and most of all, affirmative: "We must model ourselves an affirmative idea of Zero" (1993: 5). His affirmative Zero is neither self-engendering nor engendered by outside forces. Resist setting up shop among others, carving up or sifting intensities. Sometimes this is called "pagan" by Lyotard, extolling experimentation, performance, singularity, and love of the power of narrative, mixed with self-deprecation and profanity. This is not especially helpful, except for presenting the idea that signs riven by intensities are unstable and un-exchangeable, and this makes his appeal to modeling seem uncritical. While purporting to avoid nihilism and negation, Lyotard's loci of intensity are awkward to work with and their virtue may be only in that feature. Still, affirmation in the spirit of another kind of sign warrants further reflection, augmentation, and invention.

Conclusion

The shift from enthusiasm for signifiers to the generation of particle-signs marks a transition not simply from units to parts but into a dust consisting of heterogeneous segments and fragments. In the process, these particles lose the structuration provided by preset or "superior coordinate systems" (Protevi 2009: 94) and achieve an autonomy of sorts, which Guattari calls by the name of "partial enunciators" after the psychoanalytic concept of part-object inassimilable or irrecuperable, once lost (sacrificed) to a specular whole. This does not entail that these particles cannot be assembled into patterns and networks either by self-organization or by guidance, or both. No longer serving a pre-given subject, neither under the yoke of a linguistic signifier, nor beholden to a centered meaning component, sign fragments proliferate and some are selected to perform and promote (and others shine brightly in a relative isolation), and Guattari's primary examples are artistic (poetics, music, dance, theater) *molecular* ruptures of dominant significations (i.e., classification-bending) that positively feed into adventurous processes of subjectivation. The vividness of signifiers diagnosed by Jameson need not invoke schizophrenia, but, rather, appeal to aesthetic perceptual disruption as a formal strategy familiar to flicker films and countercultural uses of stroboscopes to induce distortion and at times an insufferable brilliance that surpasses representation by unleashing molecular forces (Powell 2007: 107–10).

In *Chaosmosis*, Guattari explains his transformation of Hjelmslev's categories by first breaking the parallel between expression–content and signifier–signified, and then by rejecting the tripartite hylomorphism of form–substance–matter, referring to substance as a "difficulty" to be creatively reconfigured. Taking advantage of the "reversibility" of expression and content, that is, rejecting content's positioning as an effect of primary expression, Guattari (1995: 23) advances the bidirectional pointing of the Hjelmslevian sign, to the point where even the two planes need to be stripped of a residual binarism that homogenizes and disempowers them, and any distinction between them becomes moot. He finds, then, that the so-called functives of expression and content have the virtue of reversibility based on a "common and commuting" machine that bridges the planes in relativized, unmapped, transversal ways. Guattari emphasizes passages across the planes rather than the ordering of the planes themselves; intensities escape capture by binary (expression/content) and triangular (signifier/signified/referent) abstractions. Such bridging is not restricted to language. Guattari wants to introduce a great variety of substances into the picture, but not in the way Hjelmslev imagined. After all, he thought that the expression-form and content-forms subsumed the expression-substance (sound sequences) and content-substances (thoughts of referents). Rather, Guattari imagines that enunciative substances take priority over expression–content relations and the latter, once delinguistified, can include anything from the biological to the social, especially nonlinguistic elements.

However, this opening to heterogeneous enunciative substances does not mean that Hjelmslev's tripartite form–substance–matter combination survives. Indeed, it must be "shattered" and substance moved "outside" so that "the problem of the enunciative assemblage would then no longer be specific to a semiotic register but would traverse an ensemble of heterogeneous expressive materials" (Guattari 1995: 24). Shatter and pluralize: these are Guattari's solutions to overcoming an imperial linguistics, which moves far away from sciences of pure form. And non-semiotically-formed matter must also be included, such as the sonorous materials identified in Deleuze and Guattari's (1986: 6) study of Franz Kafka, consisting of partial-signs of intensity (a-signifying sounds of all kinds issuing from animals and instruments). Not to forget a literary machine that deforms, that is, destructures both expression and content forms, as in Kafka's letters to Felice (wherein the "flux of letters" replaces the point of the correspondence, love letters that disperse love and fictionalize the recipient) (Deleuze and Guattari 1986: 28ff). Perhaps Deleuze (1997: 1) should be our guide: "to write is certainly not to impose a form (of expression) on the matter of lived experience." In place of substance Deleuze places becoming, and favors a number of attributes of a not fully determined writing: incompleteness (love of fragments), penchant for indefinite articles, stretching toward the limits of syntax (full of dashes and parentheses), production of new intensities (by means of interruptions, creative stuttering, indetermination).

Guattari's inventiveness is directed toward the preservation of intensities and multiplicities on relatively destructured semiotic strata, as noted earlier. He wants to maintain unstable, provisional states that show persistence without crashing, before succumbing to the reassertion of a stabilizing formal structure. His choice favors fragility and instability, but as active components resisting semiological recuperation. *Particle-signs* play these roles within a semiotic micropolitics that places its bets on innovation and transformation and is motivated by a politics of refusal: "the refusal of legitimating the signifying power demonstrated by the 'evidences' of dominant 'grammaticalities'" (Guattari 2011a: 176). Guattari's invention of particle-signs heralds a new semiotic life for signals, and his machinic examples ground them nonexclusively in informatics, not to forget microbials. From a Guattarian perspective, cultural studies misfires theoretically when it asks about modes of resignification (borrowing and redefinition) that involves appropriation of commodities, redefinitions of their uses and values, and their relocation in new social contexts. The theoretical error is in failing to reach beyond signification into a post-signifying universe where desubjectification offers greater molecular potential than reformed identities and subcultural symbolism. Particle-signs evoke simultaneously a-subjective and a-signifying traits, nonindividuated substances, the connection of parts to energies or what might be called a flux-y semiosis.

Anti-semiology looks suspiciously like ideology critique. Where semiolinguistics sees a science of signs, Baudrillard identifies metaphysical postulates that perfuse sign theories: he points explicitly to the naturalized use of binary oppositions, despite the unevenness of the strategic values of the signifier and economic exchange value, and the tactical values of the signified/referent and use value. Under the crushing weight of the signifier, which Guattari identified with imperial ambition, semio-linguistics erects alibis of the real in the effects of meaning and reference (signified/referent), as well as in the false objectivity identified in the illusory literalness of denotation. Baudrillard finds in the problem of reference nonexistent gaps that are papered over with weak concepts. Guattari also exposed these and other assumptions behind the scientific exterior of semiolinguistics, calling them dominant significations or realities reproduced in the study of signs in a so-called neutral manner, only by means of the exclusion of an analysis of social power. To which may be added, by the failure to point out that there is nothing self-evident and commonsensical about the relation of key binaries like expression and content: "content and expression are not attached to one another by virtue of the Holy Ghost" (2011a: 45).

While there are certainly elements of ideology critique in both the work of Baudrillard and Guattari, this does not mean that they fail to advance upon the unveiling stage and into positive and productive constructions of alternative accounts. Although Baudrillard can certainly be seen to have recourse to a master binary between semio-simulation and symbolic exchange, despite his warnings about the perils of thinking in these terms, he doesn't attempt to disguise his anthropological enthusiasm. Baudrillard may not have acknowledged, as Lyotard (1993: 106) insisted, that he reasserted "Western racism and imperialism" in its entirety in his "subversive" and "non-alienated" symbolic. Lyotard's point is that all economies are libidinal. Forces of desire and affective intensities are on both sides of the strong bar that otherwise separates semio-simulation and symbolic exchange. The paradoxical semiogram that Lyotard uncovers includes both the evacuation of intensities and the conduction of them: superimposition of signs and their others. Together, bridging strata and reducing sterility and promoting stability require greater attention to flows and variations.

Critical semiotic theory needs to take into account degrees of intensities along a continuum of formed to unformed matters (Deleuze and Guattari 1987: 511). This language is familiar to readers of Peirce's (1958) classification of "ideoscopy," in his letters to Lady Welby (8.327-29), in the categories of firstness, secondness, and thirdness. In this progressive logic of relations a quasi-independent, immediate quality enters into a relation of degree with a second (i.e., based on resistance) and a third element acts as a bridge between the other categories and itself in which the terms of relatedness are set out. Peirce's thought is amenable to a tensorial interpretation of signification since the sign's effect could be an immediate affect without any specific modality (i.e., just a vital rush), whereas a second requires either physical or mental exertion or making an effort and thus an anchored energetics is involved, while the third deals with regularities (i.e., patterns and repetitions) governing how effects and energies relate to one another. But Peirce's construction of the sign-object-interpretant configuration also engages linear and logical relations that may be downplayed by the relational-qualitative perspective, thereby foregrounding passage over position and the emergent properties of a vibrant middle. De-dichotomizing releases pure alteration as an energetics from either the terms of a dichotomy or its grounding in a specific object (Massumi 2011: 5–6). Recourse to Peirce does not offer a ready escape from representation—if the dynamic object (an external actual-real) determines the immediate object's representation of it (as an internal mental image or how we grasp the dynamic object), and the sign's generation of an interpretant also has dynamic and immediate and final facets (immediate schematic assessment— one dynamical interpretation—finally what everybody would agree upon). While triadicity guarantees the circulation of forces within semiosis, it doesn't exclude factors that either oppose or modulate these (principles of linearity, causal determination, teleology, traces of realism—the truth about the world in itself, cognitivism, abstract universals ...) and, in this respect, perhaps Lyotard is right in provocatively claiming that both or many tendencies coexist. For Guattari, recourse to Peirce solved the specific problem of the centrality of linguistics in the study of signs. For affect theory, recourse to Peirce solves the specific problem of semiotically sourcing becomings that take places within a world awash with affects.

J. S. Hutta (2015: 295) has recently argued that affect theory's "generic rejection" of semiosis, meaning, and structure, with some exceptions, requires correction. The key insight is that it matters a good deal where language's alterity is situated. If what is dynamically visceral and eventful is opposed to semiotic states, the potential for mutual influence and conditioning is limited. Hutta thinks, following Guattari's distinction between sensory and problematic affects, that "affect... not only drives discourse, but discourse also conditions affect" (2015: 298), finding that semiosis (i.e., part-signs) induces affect and affect in turn triggers semiosis (content-related). As was noted earlier in Chapter 1, small supplies of discourse are found in the pathic or affective, nondiscursive domain, despite discourse's tendency to bracket and annex intensities; indeed, pathic experience is a factor in every process of subjectivation and appears to need the discursive. Pathic intensities can work either to engage discourse in making something discernible when required or to divert discourse from its overt purposes, uses, and ends by manipulating its contents, and in so doing triggering further affects. The highly contingent interfaces between discourse (expression) and nondiscourse (content) are complex in Guattari's (1996a: 160ff) work, and suffice to say that the sensory (expression-substance that is material and nonrepresentational) and problematic (content-substance with abstract references that induce affects) affect distinction, the latter folding back on the former, recapitulates the mutual induction of affect (pathic) and semiosis (discourse). However, this distinction, which I have discussed in depth elsewhere (Genosko 2009), is deployed by Hutta in support of the insertion of semiosic power (sign-activity) into a mediating role where Guattari saw a relatively inchoate and autonomous dimension (1996a: 161). The arguments I have advanced throughout this book point to the inadequacy of unproblematically reasserting Hjelmslevian categories like expression, content, and substance, and the requirement to respect soft, partial, deformed, and energetic matters that arise with swerving around well-formed semiotic substances. By the same token, affirming soft and inchoate actualizations does not uncritically valorize an amorphous mass of pure potential out of which substances are carved and signs of which are constructed. Perhaps the clearest expression of the contrasting position, also found in Eco's work on this point, is A. J. Greimas's (1987: 19) assertion that there is a semiotic system of one kind or another all the way down to the referent and throughout all of "*signifying reality*."

In order to dislodge a somewhat intractable distinction inherited from critical theory between the material and the semiotic, and to further complexify the discussion of relatively formed substances in Chapter 3, it is worth considering how Deleuze addresses the problem of signaletic matter in *Cinema 2: The Time-Image*. He eschews the search for linguistic features (i.e., double articulation) in cinematic images, without rejecting the role language plays in cinematic utterances, yet restricting the site occupied by narration (it is not a given of images nor an effect of an underlying structure). Semiological transformations of images lose a great deal, thinks Deleuze, first and foremost movement and change, and his candid remark that "not a great deal can be done with codes" (1989: 28) reflects his commitment to a Peircean semiotics loosened from linguistic dictates. Still, his own procedures diverge from categories such as syntagmatic and paradigmatic (differentiation is concerned with shifting wholes that divide and re-cohere and specification deals with components like intervals and distinct images) introducing a rich processuality that permits him to define signaletic material: "it is a plastic mass, an a-signifying and a-syntaxic material, a material not formed linguistically even though it is not amorphous and is formed semiotically, aesthetically and pragmatically" (1989: 29). Prioritizing the image's multiple features ("sensory, kinetic, intensive, affective, rhythmic, tonal, and even verbal") situates language systems as reactions to nonlanguage materials. The semiotic priority Deleuze accords signaletics is nonlinguistic but also a-signifying. Deleuze presents a new variation on the primacy of the material over the semiotic, not by reproducing the abstractions of the semiotic, but by resisting a passive amorphousness and turning the tables on linguistics by making it "flow from" signaletics. Taken together with Guattari's turn to signal-like particle-signs, preexisting meaning is disturbed and immanence is affirmed as the anteriority of the image that is nonetheless productive: "they do not yet signify" (Rodowick 1997: 42). Critical semiotic theorists can learn *not* to experience a certain shame, of the sort cultivated by Henri Lefebvre (2003: 88), when signals (banal, trivial) are encountered in the "social text."

Within the theoretical edifices of semiolinguistics as represented by Barthes, there are lines of crossover between otherwise incommensurable positions: the anti-semiology of Baudrillard and Guattari's embrace of hybrid semiotics. The linkage is through photographic semiosis. Guattari refers to the *punctum* in his essay on Japanese photographer Keiichi Tahara, "The Faciality Machine," in *Schizoanalytic Cartographies*, but rejects the part of the theory based on biography (Barthes discusses a photograph of his mother and derives his sense of photography from his personal reflections on it). Guattari accepts, however, that the details interfere with the signification of the contexts from which the figures photographed by Tahara, specifically architect Ricardo Bofill, derive their fame, and that use of light deterritorializes the facial traits, so that the necessary reference of the portrait is not the real person who sat for it, but rather the viewer of it: "My gaze finds itself 'dragged into' the bringing into existence of Bofill ... But this appropriation turns back on me, sticks to me like a sucker... In short, I am bewitched, the evil eye is put on me, I am expropriated of my interiority" (2013: 249).

The existential transferential effect that Guattari describes is what Barthes calls *punctum*: a detail that escapes language and punctures, stings or sticks to, the viewer; this is how it attracts and holds attention, by changing the viewer. Quoting Barthes's *Camera Lucida* (1981), Guattari (2013: 250) eschews referencing the sitter's identity and significations of fame for the sake of certain traits that open up new universes of reference.

Although Baudrillard has here and there in interviews claimed to have no interest in the role of the photograph in journalism, he admits taking pleasure in attending the annual Perpignan Festival of documentary photography and its potential to surprise was highly valued by him. Although he does not discuss specific works on display there, Baudrillard (2003b) does remark that photojournalistic images like the so-called Madonna in Hell photograph from 1997, otherwise titled "Woman Grieves after Massacre in Bentalha" by Algerian war reporter Hocine Zaourar for Agence France Presse (AFP), "offer victims the mirror of their distress before dispatching the image to the 'other side' to be commercialized and consumed" as global commodities (or at least as prize-winning photos—in this case a World Press Photo of the Year—considered collectible by the Newseum alongside the cameras used to take them). In other words, this bearing witness generates a "fetishism." Indeed, he also wonders whether pictures of violence in Rwanda and Baghdad are "photography"—as opposed to images in which the subject (photographer) disappears and "something arrives from elsewhere" (Hegarty 2004: 144).

Baudrillard's sense of photography is heavily influenced by *Camera Lucida* and the punctum consisting of a foreign element that involuntarily arises in the image and touches the viewer. Baudrillard wants to discover just such a punctum in the "image" in general and suggests that this does not exist in photographs that trade poetry for realism (documenting and confirming the "mirror of events"). For Baudrillard, few if any contemporary photographs realize this ambition; Guattari finds it in the portraits of Tahara, but it is rare.

What makes this most nonlinguistic sign circulate among theorists? The existential force of the punctum, its indexicality and secondness, displaces the photograph's overt iconicity (it doesn't prove the sitter exists, it is just like him, although this hardly matters), troubling its obvious signified/referentiality and the codes of portraiture. This part-sign doesn't require meaning to exercise itself on the viewer; it doesn't require language to express itself and scrambles conventional codes. Barthes considers it to be difficult to communicate about. This indexicality is aberrant in that it turns a trace of the light that made the photograph physically possible toward the viewer, an interloper of sorts, whom it affects and does not rely on the putative object (or rather subject) of the work. Here, then, is a sharable hybrid semiotic force bearing several sign traits put into play across conceptual variations of critical semiotic theory.

Foucault's contribution to critical semiotic thought lies in a neglected dimension of a more radical thesis: the so-called hurried transition in which imprisonment became a general form of punishment after the collapse of the dreams of penal reformers in the eighteenth century. Foucault excavates the historical texts that display "astonishment" that this occurred, and his focus shifts onto how to account for change, discontinuity, and mutations of this nature. The neglected dimension refers to special semiotic characters called obstacle-signs. These signs bear the following traits:

- **i** actively nonarbitrary
- **ii** built on resemblance and proximity
- **iii** tighten the signifier (crime)/signified (punishment) relationship by means of symbolic communion
- **iv** the sign is transparent
- **v** decoding of signs is made easier
- **vi** crime's pleasures are subdued
- **vii** proliferation and diversity arrest arbitrariness creep

Obstacle-signs are, in their own right, anti-semiological insofar as they reject the principle of arbitrariness. But they are also subject to a semiotic "creep" phenomenon in which arbitrariness erodes symbolic unity by making the practice of imprisonment a matter of self-evidence. Obstacle-signs still circulate in drinking and driving campaigns and in a number of areas in which punishments analogically fit the crimes, although to speak, today, of a "thousand tiny theaters of punishment" would be an exaggeration. Nevertheless, the promotion of sobriety is a diffused effect of obstacle-signs beyond their current manifestations against drinking and driving. The target is not so much alcohol as desire and pleasure as they relate to crime and these are blocked by the ineluctable signification of obstacle-signs. This is what Deleuze and Guattari (1977: 54) call anti-production through representation: obstacle-signs screw down tight crime–punishment relations and stage and exhibit these widely as representations (what are called recordings) designed to bind desire to social norms ("crime doesn't pay"; "you will be caught and the punishment will fit the crime"; "don't drink and drive"). The link between antiproduction and social production is highlighted by obstacle-signs, as "counter-forces" (Holland 1999: 62) that block enjoyment, and in this process any excess is absorbed by the implementation of theatricalities and open books.

Using post-structural resources in order to pry open and remake, the semiotic–affect relationship generated too much tumult and not enough calm, at least as far as Lyotard's work is concerned. Any given sign produces both intensity and meaning at once in a vertigo of superimposition that is, in addition, not inscribed (contrasted with obstacle-signs) on the libidinal band. The somewhat paradoxical selection of proper names as hot points of incandescence (white light/white heat) brings Lyotard and Deleuze into conversation and realizes the strangeness of language's "other" as Derrida understood it. This strangeness is due to encountering a sign's alternate identity as an affect catcher and producer. Such a sign is punctuated by motivation and expanded by thickness. Lyotard finds thickness on the way toward designata, and his appeals to visuality and spatiality possess a heterogeneity that cannot be easily reduced to a signified content. The contrast between incandesce and crystallize may, ultimately, prove to be no more than two modalities among others, but one distinction can be taken away for further consideration: multiple rather than single energies are always at stake and local hardnesses and softnesses of manifestation must also be taken into account.

Critical semiotic theory is haunted by the dissimulating shadows of the signified. Baudrillard, Lyotard, and Lévi-Strauss point to various modifications and influences, culminating in various blooms that induct forces from the "real," including symbolic contents. The glowing signifier and the empty signified are two fields of influence that do not simply conduct energy through their failure, but display relative independence with regard to what they admit and connect with. Baudrillard exposed the dissymmetry between signifier and signified, and Lévi-Strauss and Gil posed emptiness and/or disorderliness as productive attributes, exploring the borderland of the signified/referent, just as Lyotard did in his recovery of space and sensory experience in the same zone. Deleuze's double flotation, as well as Barthes's and Jakobson's observations on the granularities of absence, reinvigorated zero as empty yet functional, and therefore significant, full of virtual potential. Signifying nothing opens a path for the transversal transmission of impulses of potentiality, and affective signs can actualize something of this potential of the cluster of particles without completely stilling its movement or unduly limiting its processes of differentiation.

References

Andersen, A. (1954) *Margarine* (London: Pergamon Press).
Aristotle. (1941) "Metaphysics," in *Basic Works*, ed. R. McKeon (New York: Random House), pp. 689–926.
Arvidsson, Adam and Colleoni, Elanor. (2012) "Value in Informational Capitalism and on the Internet," *The Information Society* 28/3: 35–50.
Assange, Julian, *et al.* (2012) *Cypherpunks: Freedom and the Future of the Internet* (New York: OR Books).
Baecker, Dirk. (2013) "Systemic Theories of Communication," in *Theories and Models of Communication*, eds. P. Cobley and P.J. Schulz (Berlin: Walter de Gruyter), pp. 85–100.
Barthes, Roland. (1990) *The Fashion System*, trans. M. Ward and R. Howard (Berkeley: University of California Press).
Barthes, Roland. (1982) *The Empire of Signs*, trans. R. Howard (New York: Hill and Wang).
Barthes, Roland. (1981) *Camera Lucida: Reflections on Photography* (London: Vintage).
Barthes, Roland. (1977) "The Third Meaning," in *Image/Music/Text*, trans. S. Heath (New York: Hill and Wang), pp. 52–68.
Barthes, Roland. (1973) *Mythologies*, trans. A. Lavers (London: Harper Collins).
Barthes, Roland. (1967) *Elements of Semiology*, trans. A. Lavers and C. Smith (New York: Hill and Wang).
Barthes, Roland. (1957) "L'opération Astra," in *Mythologies* (Paris: Editions du Seuil), pp. 44–6.
Bataille, Georges. (1991) *The Accursed Share, Vol. 1*, trans. R. Hurley (Cambridge, MA: The MIT Press).
Baudrillard, Jean. (2014) *The Divine Left: A Chronicle of the Years 1977–84*, trans. D.L. Sweet (Los Angeles: Semiotext(e)).
Baudrillard, Jean. (2011) *Telemorphosis*, trans. Drew Burk (Minneapolis, MN: Univocal).
Baudrillard, Jean. (2010) *The Agony of Power*, trans. A. Hodges (Los Angeles: Semiotetxt(e)).
Baudrillard, Jean. (2003a) *Cool Memories IV. 1995–2000*, trans. C. Turner (London: Verso).
Baudrillard, Jean. (2003b) "Le photoreportage en son miroir à VISA pour l'image," *Le Monde* (29.08).
Baudrillard, Jean. (1998) *The Consumer Society: Myths and Structures*, trans. Chris Turner (London: Sage).
Baudrillard, Jean. (1996) *The System of Objects*, trans. J. Benedict (London: Verso).

Baudrillard, Jean. (1994) *Simulacra and Simulation*, trans. S.F. Glaser (Ann Arbor: The University of Michigan Press).
Baudrillard, Jean. (1993a) *Symbolic Exchange and Death*, trans. I.H. Grant (London: Sage).
Baudrillard, Jean. (1993b) "The Revenge of the Crystal: Interview with Guy Bellavance," in *Baudrillard Live*, ed. Mike Gane (London: Routledge), pp. 50–66.
Baudrillard, Jean. (1987) "Interview," eds. T. Colless, D. Kelly, and A. Cholodenko, in *The Evil Demon of Images* (Annandale: Meglamedia and Power Institute), pp. 35–50.
Baudrillard, Jean. (1983) *Simulations*, trans. P. Foss, P. Patton, P. Beitchman (New York: Semiotext(e)).
Baudrillard, Jean. (1981) *For a Critique of the Political Economy of the Sign*, trans. C. Levin (St. Louis: Telos Press).
Bennington, Geoffrey. (1988) *Lyotard, Writing the Event* (New York: Columbia University Press).
Berardi, Franco "Bifo." (2015) *Heroes: Mass Murder and Suicide* (London: Verso).
Berardi, Franco. (2012) *The Uprising: On Poetry and Finance* (Los Angeles: Semiotext(e)).
Berardi, Franco. (2011) *After the Future*, eds. G. Genosko and Nick Thoburn (Oakland: AK Press).
Berardi, Franco. (2009) *The Soul at Work: From Alienation to Autonomy*, trans. F. Cadel and G. Mecchia (Los Angeles: Semiotext(e)).
Berardi, Franco. (2008) *Félix Guattari: Thought, Friendship, and Visionary Cartography*, trans. G. Mecchia and C. Stivale (London: Palgrave Macmillan).
Berardi, Franco. (2007) "Schizo-Economy," *Substance* 36/1: 76–85.
Berardi, Franco. (2006) Berardi, "What Is the Meaning of Autonomy Today?" *Republicart*, 1–6. http://www.republicart.net/disc/realpublicspaces/berardi01_en.htm
Bertelsen, Lone and Murphie, Andrew. (2010) "An Ethics of Everyday Infinities and Powers: Félix Guattari on Affect and the Refrain," in *The Affect Theory Reader*, eds. M. Gregg and G. Seigworth (Durham: Duke University Press), pp. 138–57.
Bloch, Ernst. (1970a) *Man on His Own*, trans. E.B. Ashton (New York: Herder and Herder).
Bloch, Ernst. (1970b) *A Philosophy of the Future*, trans. J. Cumming (New York: Herder and Herder).
Bogard, William. (2006) "Welcome to the Society of Control: The Simulation of Surveillance Revisited," in *The New Politics of Surveillance and Visibility*, eds. K.D. Haggerty and R.V. Ericson (Toronto: University of Toronto Press), pp. 55–78.
Bogard, William. (1996) *The Simulation of Surveillance: Hypercontrol in Telematic Societies* (Cambridge: Cambridge University Press).
Boldingh, J. (1969) "Research," in *Margarine* (Liverpool: Liverpool University Press), pp. 167—225.
Borch-Jacobsen, Mikkel. (1991) *Lacan: The Absolute Master* (Stanford: Stanford University Press).
Borden, Roy and Thomas, Neal. (2014) "EdgeRank," in *Encyclopedia of Social Media and Politics, Vol. 5*, eds. K. Harvey and J.G. Golson (Thousand Oaks: Sage), pp. 434–6.

Bouissac, Paul. (2010) *Saussure: A Guide for the Perplexed* (London: Continuum).
Boundas, Constantin. (1996) "Deleuze-Bergson: An Ontology of the Virtual," in *Deleuze: A Critical Reader*, ed. P. Patton (Oxford: Blackwell), pp. 81–106.
Bourassa, Alan. (2002) "Literature, Language, and the Non-Human," in *A Shock to Thought: Expression after Deleuze and Guattari*, ed. B. Massumi (New York: Routledge), pp. 60–76.
Braudel, Fernand. (1994) *A History of Civilizations*, trans. R. Mayne (London: Allen Lane/Penguin).
Brown, Steven D. and Tucker, Ian. (2010) "Eff the Ineffable: Affect, Somatic Management, and Mental Health Service Users," in *The Affect Theory Reader*, eds. M. Gregg and G. Seigworth (Durham: Duke University Press), pp. 229–49.
Brunton, Finn. (2013) *Spam: A Shadow History of the Internet* (Cambridge, MA: The MIT Press).
Busch, Lawrence. (2011) *Standards* (Cambridge, MA: The MIT Press).
Cherry, Colin. (1966) *On Human Communication*, 2nd Edition (Cambridge, MA: The MIT Press).
Clough, Patricia. (2007) "Introduction," in *The Affective Turn*, eds. P. Clough and J. Halley (Durham: Duke University Press), pp. 1–33.
Cohen, Stanley. (1979) "The Punitive City: Notes on the Dispersal of Social Control," *Contemporary Crisis* 3/4: 339–63.
Coté, Mark and Pybus, Jennifer. (2007) "Learning to Immaterial Labour 2.0: MySpace and Social Networks," *ephemera* 7/1: 88–106.
Coulter, Gerry. (2012) *Jean Baudrillard: From the Ocean to the Desert, or the Poetics of Radicality* (Skyland, NC: Inter-Theory).
Daniel, J. Owen and Moylan, Tom (eds.). (1997) *Not Yet: Reconsidering Ernst Bloch* (London: Verso).
Davison, Patrick. (2012) "The Language of Internet Memes," in *The Social Media Reader*, ed. M. Mandiberg (New York: New York University Press), pp. 120–34.
De Saussure, Ferdinand. (2011) *Course in General Linguistics*, trans. W. Baskin, eds. P. Meisel and H. Saussy (New York: Columbia University Press).
Deleuze, Gilles. (2003) *Francis Bacon: The Logic of Sensation*, trans. D. W. Smith (Minneapolis: University of Minnesota Press).
Deleuze, Gilles. (1997) "Literature and Life," in *Essays Critical and Clinical*, trans. Daniel Smith and Michael Greco (Minneapolis: University of Minnesota Press), pp. 1–6.
Deleuze, Gilles. (1994) *Difference and Repetition*, trans. Paul Patton (New York: Columbia University Press).
Deleuze, Gilles. (1992) "Postscript on Societies of Control," *October* 59 (Winter): 3–7.
Deleuze, Gilles. (1990) *The Logic of Sense*, trans. Mark Lester and Charles Stivale, ed. Constantin Boundas (New York: Columbia University Press).
Deleuze, Gilles. (1989) *Cinema 2: The Time-Image*, trans. H. Tomlinson and R. Galeta (Minneapolis: University of Minesota Press).
Deleuze, Gilles. (1988) *Spinoza: Practical Philosophy*, trans. R. Hurley (San Francisco: City Lights).
Deleuze, Gilles and Guattari, Félix. (1987) *A Thousand Plateaus*, trans. B. Massumi (Minneapolis: University of Minnesota Press).

Deleuze, Gilles and Guattari, Félix. (1986) *Kafka: For a Minor Literature*, trans. D. Polan (Minneaplois: University of Minnesota Press).
Deleuze, Gilles and Guattari, Félix. (1977) *Anti-Oedipus*, trans. R. Hurley, Mark Seem, H.R. Lane (New York: The Viking Press).
Derrida, Jacques. (1992) *Given Time: I. Counterfeit Money*, trans. P. Kamuf (Chicago: University of Chicago Press).
Derrida, Jacques. (1981) *Positions,* trans. Alan Bass (Chicago: University of Chicago Press).
Derrida, Jacques. (1978) *Writing and Difference*, trans. Alan Bass (Chicago: The University of Chicago Press).
Derrida, Jacques. (1973) *Speech and Phenomena*, trans. D.B. Allison (Evanston: Northwestern University Press).
Dosse, François. (2012) "Deleuze and Structuralism," in *The Cambridge Companion to Deleuze*, eds. D.W. Smith and H. Somers-Hall (Cambridge: Cambridge University Press), pp. 126–50.
Dyer-Withford, Nick and De Peuter, Greig. (2009) *Games of Empire: Global Capitalism and Video Games* (Minneapolis: University of Minnesota Press).
Eco, Umberto, Eco. (1982) "On Symbols," *Recherches Sémiotiques/Semiotic Inquiry* 2/1: 15–44.
Eco, Umberto, Eco. (1976) *A Theory of Semiotics* (Bloomington: Indiana University Press).
Elmer, Greg. (2004) *Profiling Machines: Mapping the Personal Information Economy* (Cambridge: The MIT Press).
Feron, R. (1969) "Technology and Production," in *Margarine: An Economic, Social, and Scientific History 1869–1969*, ed. J.H. van Stuyvenberg (Liverpool: Liverpool University Press), pp. 83–121.
Fortunati, Leopoldina. (2007) "Immaterial Labour and Its Machinization," *ephemera* 7/1: 139–57.
Foucault, Michel. (2002) "'Space, Knowledge, and Power' and 'Questions of Method'," in *Essential Works*, Vol. 3: Power, trans. R. Hurley, et al. (London: Penguin), pp. 349–64 and pp. 223–38.
Foucault, Michel. (1977) *Discipline and Punish: The Birth of the Prison*, trans. Alan Sheridan (New York: Random House).
Foucault, Michel. (1975) *Surveiller et punir: Naissance de la prison* (Paris: Gallimard).
Foucault, Michel. (1972) *The Archaeology of Knowledge*, trans. A.M. Sheridan Smith (New York: Pantheon).
Freud, Sigmund. (2001) "The Interpretation of Dreams," in *The Standard Edition of the Complete Psychoanalytical Works*, Vols. IV–V, trans. James Strachey (London: Vintage, 2001), pp. iii–338; iii–626.
Furján, Helene. (2008) "On Eco-Logics," *Artforum* (November): 295–7 and 374.
Gabrys, Jennifer. (2010) "Telepathically Urban," in *Circulation and the City: Essays on Urban Culture*, eds. A. Boutros and W. Straw (Montréal: McGill-Queen's University Press), pp. 48–63.
Galloway, Alexander R. (2012) *The Interface Effect* (Cambridge: Polity Press).
Gane, Mike. (1991) *Baudrillard: Critical and Fatal Theory* (London: Routledge).
Gasparov, Boris. (2013) *Beyond Pure Reason: Ferdinand de Saussure's Philosophy of Language and Its Early Romantic Antecedents* (New York: Columbia University Press).

Genosko, Gary. (2015) "Pathic Transferences and Contemporary Japanese Art," in *Félix Guattari, Machinic Eros: Writings on Japan*, eds. G. Genosko and Jay Hetrick (Minneapolis: Univocal), pp. 97–115.
Genosko, Gary. (2012) *Remodelling Communication: From WWII to the WWW* (Toronto: University of Toronto Press).
Genosko, Gary. (2010) "The Terrorist Entrepreneur," *TOPIA* 23–24 (Special Issue on Cultures of Militarization): 330–44.
Genosko, Gary. (2009) *Félix Guattari: A Critical Introduction* (London: Pluto).
Genosko, Gary. (2002) *Félix Guattari: An Aberrant Introduction* (London: Continuum).
Genosko, Gary. (1998) *Undisciplined Theory* (London: Sage).
Genosko, Gary. (1994) *Baudrillard and Signs* (London: Routledge).
Gil, Jose. (1990) *Metamorphoses of the Body*, trans. S. Muecke (Minneapolis: University of Minnesota Press).
Gleick, James. (2011) *The Information: A History, A Theory, A Flood* (New York: Pantheon).
Grace, Victoria. (2000) *Baudrillard's Challenge: A Feminist Reading* (London: Routledge).
Graeber, David. (2008) "The Sadness of Post-Workerism or Art and Immaterial Labour Conference, a Sort of Review." http://www.thedepartmentofaesthetics.org/graebner_sadness.pdf
Greimas, A.J. (1987) *On Meaning: Selected Writings in Semiotic Theory*, trans, P. Perron and F. Collins (Minneapolis: University of Minnesota Press).
Grosz, Elizabeth. (2008) *Chaos, Territory, Art: Deleuze and the Framing of the Earth* (New York: Columbia University Press).
Guattari, Félix. (2015) "Ecosophical Practices and the Restoration of the 'Subjective City,'" in *Félix Guattari, Machinic Eros: Writings on Japan*, trans. G. Genosko (Minneapolis: Univocal), pp. 97–115.
Guattari, Félix. (2013) *Schizoanalytic Cartographies*, trans. A Goffey (London: Bloomsbury).
Guattari, Félix. (2011a) *The Machinic Unconscious*, trans. T. Adkins (Los Angeles: Semiotext(e)).
Guattari, Félix. (2011b) *Lignes de fuite: Pour un autre monde de possibles* (La Tour d'Aigues: Editions de l'aube).
Guattari, Félix. (2011c) "The Vertigo of Immanence," in *The Guattari Effect*, eds. E. Alliez and A. Goffey (London: Continuum), pp. 25–39.
Guattari, Félix. (2008) *The Three Ecologies*, trans. I. Pindar and P. Sutton (London: Bloomsbury).
Guattari, Félix. (1998) "La grille," *Chimères* 34: 7–20.
Guattari, Félix. (1996a) *The Guattari Reader*, ed. G. Genosko (Oxford: Blackwell).
Guattari, Félix. (1996b) *Soft Subversions* (New York: Semiotext(e)), pp. 202–24.
Guattari, Félix. (1995) *Chaosmosis*, trans. P. Bains and J. Pefanis (Bloomington: Indiana University Press).
Guattari, Félix. (1993) "Space and Corporeity," *D: Columbia Documents in Architecture and Theory* 2: 139–49.
Guattari, Félix. (1990) "La révolution moléculaire," *Le Monde* 7/12: 2.
Guattari, Félix. (1986) "Questionnaire on the City," trans. B. Benderson, *Zone* 1/2: 460.

Guattari, Félix. (1981) "Quelle est pour vous la Cité idéale?," *La Quinzaine littéraire* 353: 39.
Guattari, Félix. (1978) *La révolution moléculaire* (Paris: Edition 10–18).
Guattari, Félix. (1977) *La révolution moléculaire* (Fontenay-sous-Bois: Encres/ Recherches).
Hardt, Michael. (2006) "Production and Distribution of the Common," in *Being an Artist in Post-Fordist Times*, eds. P. Gielen and P. De Bruyne (Rotterdam: NAI), pp. 43–53.
Hardt, Michael and Negri, Toni. (2009) *Commonwealth* (Cambridge, MA: Belknap Press).
Hardt, Michael and Negri, Toni. (2004) *Multitude: War and Democracy in the Age of Empire* (New York: Penguin).
Hardt, Michael and Negri, Toni. (2000) *Empire* (Cambridge, MA: Harvard University Press).
Harris, Roy. (1988) *Language, Saussure and Wittgenstein: How to Play Games with Words* (London: Routledge).
Hegarty, Paul. (2004) *Baudrillard Live* (London: Continuum).
Heron, Craig. (2003) *Booze: A Distilled History* (Toronto: Between the Lines).
Hjelmslev, Louis. (1969) *Prolegomena to a Theory of Language*, trans. Francis J. Whitfield (Madison: University of Wisconsin Press).
Holland, Eugene (1999) *Deleuze and Guattari's Anti-Oedipus: Introduction to Schizoanalysis* (New York: Routledge).
Holmes, Brian. (2009) "Guattari's Schizoanalytic Cartographies or the Pathic Core at the Heart of Cybernetics," in *Continental Drift* Blog (02/27). https://brianholmes.wordpress.com/2009/02/27/guattaris-schizoanalytic-cartographies/
Hudson, Wayne. (1982) *The Marxist Philosophy of Ernst Bloch* (New York: St. Martin's Press).
Hutta, J.S. (2015) "The Affective Life of Semiotics," *Geographica Helvetica* 70: 295–309.
Jakobson, Roman. (1971) "Notes on the French Phonemic Pattern," in *Selected Writings: Phonological Studies* (The Hague: Mouton de Gruyter), pp. 426–34.
Jameson, Fredric. (1991) *Postmodernism, or, The Cultural Logic of Late Capitalism* (Durham: Duke University Press).
Jameson, Fredric. (1984) "Periodizing the Sixties," *Social Text* 9/10 (Summer): 172–209.
Kellner, Douglas. (1989) "Jameson/Marxism/Postmodernism," in *Postmodernism/ Jameson/Critique* (Washington: Maisonneuve Press), pp. 1–42.
Kristeva, Julia (1986) "The System and the Speaking Subject," in *The Kristeva Reader*, ed. Toil Moi (Oxford: Blackwell), pp. 24–33.
Lacan, Jacques. (1977) *Écrits: A Selection*, trans. A. Sheridan (New York: W.W. Norton).
Langlois, Ganaele. (2014) *Meaning in the Age of Social Media* (New York: Palgrave Macmillan).
Lazzarato, Maurizio. (2014) *Signs and Machines: Capitalism and the Production of Subjectivity*, trans. J.D. Jordan (Los Angeles: Semiottext(e)).
Lazzarato, Maurizio. (2011) *The Making of Indebted Man*, trans. J.D. Jordan (Los Angeles: Semiotext(e)).

Lazzarato, Maurizio. (2006a) "The machine," trans. Mary O'Neill, European Institute for Progressive Cultural Politics (EIPCP Transversal Texts). http://www.eipcp.net/transversal/1106/lazzarato/en

Lazzarato, Maurizio. (2006b) "'Semiotic Pluralism' and the New Government of Signs: Homage to Félix Guattari," European Institute for Progressive Cultural Politics (EIPCP Transversal Texts). http://translate.eipcp.net/transversal/0107/lazzarato/en#redir

Lazzarato, Maurizio. (1996) "Immaterial Labour," in *Radical Thought in Italy: A Potential Politics*, eds. Paolo Virno and Michael Hardt (Minneapolis: University of Minnesota Press), pp. 132–46.

Leary, Timothy. (1968) *The Politics of Ecstasy* (New York: College Notes and Texts).

Lefebvre, Henri. (2003) "The Social Text," in *Key Writings* (London: Continuum), pp. 88–92.

Lerner, Barron H. (2011) *One for the Road: Drunk Driving since 1900* (Baltimore, MD: Johns Hopkins University Press).

Lévi-Strauss, Claude. (1987) *Introduction to the Work of Marcel Mauss*, trans. Felicity Beck (London: Routledge & Kegan Paul).

Lévi-Strauss, Claude. (1963) *Structural Anthropology*, trans. C. Jacobson and B.G. Schoepf (New York: Doubleday Anchor).

Lyotard, Jean-François. (2011) *Discourse, Figure*, trans. A. Hudek and M. Lydon (Minneapolis: University of Minnesota Press).

Lyotard, Jean-François. (1993) *Libidinal Economy*, trans. I.H. Grant (Bloomington: Indiana University Press).

Maniglier, Patrice. (2010) *The History of Continental Philosophy, Volume 7. After Poststructuralism: Transitions and Transformations*, ed. Rosi Braidotti (Durham: Acumen), pp. 55–82.

Mann, Thomas. (1996) *The Magic Mountain* (New York: Vintage).

Maras, Steven. (2008) '"On Transmission: A Metamethodological Analysis," *Fibreculture* 12: http://twelve.fibreculturejournal.org/fcj-080-on-transmission-a-metamethodological-analysis-after-regis-debray/

Marazzi, Christian. (2010) *The Violence of Financial Capitalism*, trans. K. Lebedeva (Los Angeles: Semiotext(e)).

Marcus, Anthony. (2005) '"Drinking Politics: Alcohol, Drugs and the Problem of US Civil Society," in *Drinking Cultures*, ed. T.A. Wilson (Oxford: Berg), pp. 255–67.

Marx, Karl. (1972a) *A Contribution to the Critique of Political Economy*, trans. S.W. Ryazanskaya (New York: International Publishers).

Marx, Karl. (1972b) *Grundrisse*, trans. Martin Nicolaus (New York: Vintage).

Marx, Karl. (1954) *Capital 1*, trans. S. Moore and E. Aveling (Moscow: Progress).

Massumi, Brian. (2011) *Semblance and Event* (Cambridge: The MIT Press).

Massumi, Brian. (2002) *Parables for the Virtual* (Durham: Duke University Press).

Massumi, Brian. (1992) *A User's Guide to Capitalism and Schizophrenia: Deviations from Deleuze and Guattari* (Cambridge: The MIT Press).

Mauss, Marcel. (1990) *The Gift*, trans. W.D. Halls (London: Routledge).

Mehlmann, Jeffrey. (1972) "The 'Floating Signifier': From Lévi-Strauss to Lacan," *Yale French Studies* 48: 10–37.

Melitopoulos, Angela and Lazzarato, Maurizio. (2012) "Machinic Animism," *Deleuze Studies* 6/2: 92–101.

Mumford, Lewis. (1970) *The Myth of the Machine: The Pentagon of Power* (New York: Harcourt Brace Jovanovich).
Mumford, Lewis. (1961) *The City in History: Its Orgins, Its Transfromations, and Its Prospects* (New York: Harcourt, Brace and World).
Negarestani, Rez. (2008) *Cyclonopedia: Complicity with Anonymous Materials* (Melbourne: Re:press).
Negri, Antonio. (2008a) *The Porcelain Workshop: For a New Grammar of Politics*, trans. N. Wedell (Los Angeles: Semiotext(e)).
Negri, Antonio. (2008b) *Reflections on Empire*, trans. Ed Emery (Cambridge: Polity).
Negri, Antonio. (2008c) *Empire and Beyond*, trans. Ed Emery (Cambridge: Polity).
Negri, Antonio. (2003) *Time for Revolution*, trans. M. Mandarini (New York: Continuum).
Ogden, C.K. and Richards, I.A. (1923) *The Meaning of Meaning* (New York: Harcourt, Brace & World).
Pasquinelli, Matteo. (2009) "Google's PageRank Algorithm: A Diagram of Cognitive Capitalism and the Rentier of the Common Intellect," in *Deep Search*, eds. Konrad Becker and Felix Stalder (London: Transaction Publishers). http://matteopasquinelli.com/google-pagerank-algorithm/
Peirce, C.S. (1958) *Collected Papers of Charles Sanders Peirce*, Vol. VIII, ed. A.W. Burks (Cambridge: Harvard University Press).
Peirce, C.S. (1932) *Collected Papers of Charles Sanders Peirce*, Vol. II, eds. C. Hartshorne and P. Weiss (Cambridge: Harvard University Press).
Powell, Anna. (2007) *Deleuze, Altered States and Film* (Edinburgh: Edinburgh University Press).
Protevi, John. (2009) *Political Affect: Connecting the Social and the Somatic* (Minneapolis: University of Minnesota Press).
Protevi, John. (2001) *Political Physics: Deleuze, Derrida and the Body Politic* (London: The Athlone Press).
Querrien, Anne. (2011) "Maps and Refrains of a Rainbow Panther," in *The Guattari Effect*, trans. G. Genosko (London: Continuum), pp. 84–98.
Ransom, John S. (1997) *Foucault's Discipline: The Politics of Subjectivity* (Durham, NC: Duke University Press).
Raunig, Gerald. (2013) *Factories of Knowledge, Industries of Creativity* (Los Angeles: Semiotext(e)).
Readings, Bill. (1991) *Introducing Lyotard: Art and Politics* (New York: Routledge, 1991).
Rodowick, David N. (1997) *Gilles Deleuze's Time Machine* (Durham: Duke University Press).
Rossi-Landi, F. (1977) *Linguistics and Semiotics* (The Hague: Mouton).
Rotman, Brian. (1987) *Signifying Nothing: The Semiotics of Zero* (New York: St. Martin's Press).
Ruoff, Jeffrey. (2001) *An American Family: A Televised Life* (Minneapolis: University of Minnesota Press).
Rynne, Colin. (1998) *At the Sign of the Cow: The Cork Butter Market 1770–1969* (Cork: Collins).
Schirato, Tony. (1998) "Meaning," in *Encyclopedia of Semiotics*, ed. P. Bouissac (Oxford: Oxford University Press), pp. 396–99.

Seising, Rudolf. (2009) "60 years of 'A Mathematical Theory of Communication' – Towards a 'Fuzzy Information Theory,'" *IFSA-EUSFLAT*: 1332–37.
Shannon, Claude. (1948) "A Mathematical Theory of Communication," *Bell System Technical Journal*, 27 (July and October): 379–423; 623–56.
Sim, Stuart. (2011) "Great Zero," in *The Lyotard Dictionary*, ed. S. Sim (Edinburgh: Edinburgh University Press), pp. 91–2.
Simpkins, Scott. (2001) *Literary Semiotics: A Critical Approach* (Lanham, MD: Lexington Books).
Skyrms, B. (2010) *Signals* (Oxford: Oxford University Press).
Smith, Daniel W. (1996) "Deleuze's Theory of Sensation: Overcoming the Kantian Duality," in *Deleuze: A Critical Reader*, ed. Paul Patton (Oxford: Blackwell), pp. 29–56.
Steyerl, Hito. (2014) "Proxy Politics: Signal and Noise," *E-flux* #60 (December). http://www.e-flux.com/journal/proxy-politics/#_ftn20
Van Stuyvenberg, J.H. (1969) "Aspects of Government Intervention," in *Margarine*, ed. J.H. van Stuyvenberg (Liverpool: Liverpool University Press), pp. 281–327.
Terranova, Tiziana. (2004) *Network Culture: Politics for the Information Age* (London: Pluto).
Thacker, Eugene. (2011) *In the Dust of This Planet: Horror of Philosophy, Vol. 1* (Alresford, Hants: Zero Books).
Toscano, Alberto. (2007) "Vital Strategies: Maurizio Lazzarato and the Metaphysics of Contemporary Capitalism," *Theory, Culture & Society* 24/6: 71–91.
Virno, Paolo. (2006) "The Dismeasure of Art: An Interview with Paolo Virno," in *Being an Artist in Post-Fordist Times*, eds. P. Gielen and P. De Bruyne (Rotterdam: NAI), pp. 17–44.
Virno, Paolo. (2001) "General Intellect," trans. A. Bove, *Generation Online* http://generation-online.org/p/fpvirno10.htm
Virtanen, Askeli. (2004) "General Economy: The Entrance of the Multitude into Production," *ephemera* 4/3: 209–32.
Visker, Rudi. (1995) *Michel Foucault: Geneaology as Critique*, trans. C. Turner (London: Verso).
Visser, Margaret. (1986) *Much Depends on Dinner* (Toronto: Harper Collins).
Watson, Janell. (2009) *Guattari's Diagrammatic Thought: Writing between Lacan and Deleuze* (New York: Continuum).
Weaver, Warren. (1949) "Recent Contributions to the Mathematical Theory of Communication," in Claude Shannon and Warren Weaver, *The Mathematical Theory of Communication* (Urbana: University of Illinois Press), pp. 95–117.
Welchman, Alistair. (1997) "Machinic Thinking," in *Deleuze and Philosophy: The Difference Engineer*, ed. K.A. Pearson (London: Routledge), pp. 211–29.
Wiener, Norbert. (1967) *The Human Use of Human Beings: Cybernetics and Society* (New York: Avon Books).
Wilden, Anthony. (1968) "Lacan and the Discourse of the Other," in Jacques Lacan, *Speech and Language in Psychoanalysis*, trans. A. Wilden (Baltimore: The Johns Hopkins University Press), pp. 159–311.

Index

www.ingramcontent.com/pod-product-compliance
Ingram Content Group UK Ltd.
Pitfield, Milton Keynes, MK11 3LW, UK
UKHW020143290926
15471UKWH00004B/673

9 781472 596369